Pakistan's Energy Issues: Success & Challenges

Syed Akhtar Ali

Research on Economy & Politics

(REAPP-Islamabad)

This study has been done under the auspices of REAPP- Research on Economy and Politics of Pakistan. REAPP founded in 1981, has by now commissioned many studies which have been published including the ones authored by Syed Akhtar Ali.

@copy right by the Author

Table of Contents

PART I: GENERAL 14
1. INTRODUCTION 15
2. STRATEGIC DIRECTIONS IN ENERGY SECTOR 48
3. WHAT IS THE RIGHT ENERGY SOURCE? 56
4. ACHIEVING UNIVERSAL ENERGY ACCESS IN PAKISTAN 68
5. ARE ENERGY PRICES HIGH IN PAKISTAN 75
6. ENERGY PRICES AND ECONOMICS IN PAKISTAN 81
7. WHERE WILL OIL AND ENERGY PRICES GO? 90
8. PRIORITIES IN CPEC 94
9. THE CONUNDRUM OF SINGLE BID 97

PART II: RENEWABLE ENERGY, CONSERVATION AND CLIMATE CHANGE 104
10. ARE WE CONSUMING ENERGY EFFICIENTLY? 105
11. TOWARDS A REALISTIC CLIMATE CHANGE POLICY 110
12. COMPETITION IN RENEWABLE ENERGY 114
13. BHASHA DAM: BEG, BORROW OR STEAL? 123
14. HYDRO ROYALTY 131

PART III: OIL & GAS 136
15. RESTRUCTURING OIL AND GAS SECTOR 137
16. NATURAL GAS AND LNG DEMAND 142
17. NON-CONVENTIONAL GAS RESOURCES 152
18. ALTERNATIVE APPROACHES IN RLNG 159
19. ALTERNATIVE APPROACHES TO LPG AIR MIX PLANTS: SMALL SCALE LNG SUPPLY CHAIN 164
20. LPG SUBSIDY SCHEME FOR AVERTING DEFORESTATION 173
21. LNG CONTROVERSY 177

22.	CURTAILING LOSSES IN GAS SECTOR	183
23.	PROMOTING OIL & GAS PRODUCTION	187
24.	ENFORCING ENVIRONMENTAL FUEL STANDARDS TO GET RID OF SMOG	190

PART IV: COAL 193

25.	THE PROBLEMS AND PROSPECTS OF COAL?	194

SOURCE: NEPRA TARIFF DETERMINATION 202

26.	NEW THAR COAL TARIFF	207
27.	ENVIRONMENT AND COAL POWER	217
28.	UNDERGROUND COAL GASIFICATION	224
29.	INDUSTRIAL AND AGRICULTURAL USES OF THAR COAL	227

PART IV: ELECTRICITY 231

30.	HOW MUCH ELECTRICITY DO WE NEED?	232
31.	LOAD SHEDDING AND TRANSPARENCY	246
32.	DISCO PERFORMANCE	252
33.	REORGANIZATION OF THE POWER SECTOR	261
34.	FEDERALISM IN ELECTRICITY	265
35.	REVIVING PEPCO	275
36.	REGULATORY REFORMS	278
37.	ALTERNATIVE APPROACHES TO SMART METERING	286
38.	CASA - A LIABILITY?	289
39.	NEW K-ELECTRIC TARIFF	294
40.	RELENDING CHARGES IN THE ENERGY SECTOR	297
41.	NUCLEAR POWER	300
42.	DEMYSTIFYING NANDIPUR PROJECT CONTROVERSY	305

List of Tables

Table 1 Energy Components Growth Rates (%p.a.) 49

Table 2 Pakistan's Power Generation Mix 2015-16 51

Table 3 Electric Power Installed Capacity vs. Water Storage 59

Table 4 Comparative parameters of Various Energy Sources 66

Table 5 Comparison of Electricity Access 2016 69

Table 6 Health Effects of Hazardous Air Pollution 73

Table 7 Gasoline and Diesel Prices in Selected Countries (May 22-2017) 76

Table 8 Comparative Electricity Tariffs in India and Pakistan (USc per kWh) 78

Table 9 Gas & Electricity Prices for Household and Industry in Europe (2016-2017) 80

Table 10 Comparative Fuel Cost in Electricity Production (Rs/kWh) 82

Table 11 Comparative Coal Prices 84

Table 12: Comparative Tariff: Thar vs. Imported Coal 85

Table 13 Cost of Generation (COGE) of RLNG NGCC vs. Imported Coal 86

Table 14 LNG Supply Contracts and Prices 87

Table 15 Natural Gas Prices in Pakistan (USD/MMBtu) 89

Table 16 Energy Prices Projections (2015-2040) at nominal and constant prices of 2015 in USD 93

Table 17 Bilateral Project Loan Terms in South Asia 103

Table 18 Comparative INDC Targets of Various Countries 111

Table 19 Wind Power Levellised Tariff in 5 Wind Zones in India (FY2016-17) 120

Table 20 Wind Power Tariff Parameters: GERC India vs NEPRA 121

Table 21 Comparative Royalty Rates in Hydro sector in Various Countries 133

Table 22 Local Oil and Gas Production Trend (2002-15) 143

Table 23 Natural Gas Consumption (2002-2015) 145

Table 24 Natural Gas Consumption Forecast (2015-2025) 146

Table 25 Natural Gas: Projected Demand and Supply (mmcfd) 2025 151

Table 26 Small Scale LNG(SSLNG) perspective data and parameters 167

Table 27 Comparative LPG and HSD Prices (Whole Sale/Ex-Refinery) 168

Table 28: Proposed LPG Subsidy scheme-Major data and parameters 176

Table 29: Comparative LNG Import Prices in Countries 180

Table 30: RLNG Prices Sept 2017(USD per MMBtu) 181

Table 31 Actual UFG vs. Allowed UFG in SSGC and SNGPL over the recent years 185

Table 32 Environmental Motor Fuel Standards in India 192

Table 33 Capital Costs for Construction of New Coal Fired Power Plants (US$mn) 196

Table 34: China CAPEX Estimates Coal Power Plants of various capacities 197

Table 35 Summary of NEPRA Tariff: PQEPP Coal Power Plant at Port Qasim/Sahiwal ... 202

Table 36 Comparative Tariff: Thar vs. Imported Coal 211

Table 37: Comparative Tariff Determinations of Thar Coal Mining Projects 216

Table 38: WHO Guidelines for Ambient Air Quality (mg/M^3) 221

Table 39 Emission Standards of Coal Power Plants: New vs. Old and Comparative Environmental standards of various power plants (mg/M3) 222

Table 40 Salient Data on Economies and Electricity consumption of Selected Countries-2014 235

Table 41 Comparison of JICA and SNC-LAVALIN Studies: Peak Demand Forecasts (MW) 238

Table 42 SNC-Lavalin Study: Electrical Peak Demand Forecast 2035 240

Table 43 JICA Study: Electricity Peak Demand Forecast up to 2035 240

Table 44 Demand and Supply Projections up to 2035 241

Table 45 Summary of New Additions and Net Total Capacity by 2035 243

Table 46 List of Electrical Power Projects 243

Table 47 Aggregated Capacity Utilization Factors 248

Table 48 Major Power Plants and their Utilization in 2015-16 249

Table 49 Salient Parameters of Distribution Companies..260

Table 50 Pakistan Interest Rates March-May 2017..299

Table 51 Comparative Nuclear Power Cost from Western Vendors..............................302

Table 52 Nuclear Power Cost in Pakistan 302

Table 53 Nandipur Turbine Specifications...306

Table 54 Nandipur Approved v Claimed Cost..306

Table 55 Comparison with other IPPs 308

Table 56 Comparison with other Gas Plants 308

List of Figures

Figure 1 Future Demand and Supply 52

Figure 2 Future Demand and Supply 2014-35 53

Figure 3 Power Development Plan up to 2035: Demand and Supply Balance (MW) 53

Figure 4 Village Electrification in Distribution Companies 70

Figure 5 Percentage of Population using Traditional Sources of Fuel..............................72

Figure 6 Break Even Prices of Oil Companies 93

Figure 7 Energy Intensity (GDP per kg of Oil) of Selected Countries..............................106

Figure 8 Pakistan Energy Label 108

Figure 9 Emissions Intensity (per GDP-PPP) 110

Figure 10 Average Prices Resulting from Auctions, 2010-16 116

Figure 11 Wind Turbine Bidding Prices Stable 119

Figure 12 Local Oil and Gas Production Trend (2002-15)................................144

Figure 14 LCNG Stations 172

Figure 15 Processes of Pulverized Lignite Burner..230

Figure 16 Demand and Supply Situation January to November 2015 247

Figure 17 DISCOs T&D Losses (%) 254

Figure 18 Recovery Rate of DISCOs (%) 255

Figure 19 System Average Interruption Duration Index (SAIDI).......................256

Figure 20 System Average Interruption Frequency (SAIFI) 257

Preface

I wrote two books almost on the same subject, but with varying emphasis and focus, in the recent past: Pakistan's Energy Development: The Road Ahead (2009) and; Issues in Energy Policy (2013).The first one dealt with some technical issues as well besides policy and economics and the second one dealt with policy issues only. Why yet another book so early after? CPEC and Oil price reduction have changed energy scenario significantly, which has opened up new opportunities and have brought forward many issues and problems. This book deals with the period of 2014 and onwards, although there are inter-relationships among the periods.

A friend of mine quipped as to why I am writing all this and issuing all kind of suggestions after leaving the job of Member Energy at Planning Commission and that the advice should have been given directly within government mechanism. First of all, there is no free and candid mechanism in GoP to discuss these issues except for the comments and summaries and the same have been used by this scribe and has been quite verbal about in corridor talks. Some have been accepted and others have not been. The issues that have been dealt with in most part of the book deserve to be discussed in public domain, for the future policy and decision-makers may emerge out of any quarter. And the public should know principle.

There is no agenda of this book; the aim is to present facts, analysis, conclusion and recommendations, as I deem appropriate and useful. There are positives and negatives. We are still mid-way through and something can be done to control the negatives. In the period that is the focus of the book, it was an extremely fast track process to solve the crisis. The energy crisis appears to have been fixed. The issue is what next. Where and how do we go from here? How much is enough? Can we afford excess? What is the cost of excess and benefit of abundance? What are the options and choices?

There are many issues that are being debated among policy and stakeholders circle; Are energy prices high in Pakistan? Are these going to go down with the new projects? What is the right source of Energy for us? Should we adopt Renewable Energy as has been done and being done by Europe or we have constraints? What would be the impact of future oil and gas prices? Are these going to remain low as today or are going to repeat the earlier cycle? Is imported coal a transition solution and we should revert to Thar coal? The world is going against coal, why should we go against the current? Can there be something like clean coal? Are our coal promoters and investors abiding by the environment rules and the undertakings of EIA or are perverting the EIA process itself as happens in developing countries usually? Is LNG/RLNG the future under the low gas prices? Why can't we develop our own gas resources, if not oil, and what are the implications of imported gas? Are we going to repent by adopting a gas regime in the same way that we are repenting with Oil? And the case of small-scale LNG vs LPG, the latter being much more expensive than the former; MPNR has adopted LPG to be distributed in the new isolated networks. We have advised adoption of LNG by Trucks. Hydro Power is on our national agenda for a long time, why could we not implement it and what is the way ahead? What are the constraints? Is nuclear dead? We have selected issues and dealt with these as separate chapters. We are not sure that all answers have been dealt with adequately but at least some have been done.

I must thank many organizations whose data, publications and reports have been utilized in preparing this book, these are: NEPRA, OGRA, NTDC, PPIB, AEDB, WAPDA, HDIP, State Bank of Pakistan, and others whose name may not have been mentioned here by unintentional omission, without the useful data made available by these organization, it is impossible to write such a book as this is. I have particularly benefited from participating in the public hearings organized by NEPRA for NEPRA determinations. To Woodrow Wilson Centre, I am grateful to have invited me to speak in one of their seminars which gave me an opportunity to discuss issues with a number of leading experts which have found ways into some of the recommendations made here. UNDPs project on 'Energy Access for All' gave me the opportunity to speak to a wide variety of audience in various parts of Pakistan.

I would be remiss in my duties if I do not mention my colleagues of the Energy Wing in their role in sharpening my understanding in some of the controversial issues. I must also thank my colleagues (members) at Planning Commission Dr. NaeemuZZafar, who is now senior advisor at UNDP Islamabad and Dr. Faheem-up-Islam .Both of these gentlemen reviewed the manuscript and provided useful suggestions for improvement. Dr. Afreen Siddiqi of MIT/Harvard is also thanked for providing input on some chapters of the books. Ameena Sohail deserves acknowledgement for her review. Her knowledge of regulatory law was a good resource for me. She also provided me ample opportunities to present and test my thoughts and ideas at a number of seminars that had been organized at Institute of Policy studies Islamabad Policy which she is the board member of for giving me frequent opportunities to speak in their seminars which did help in identifying many is. I must also thank Tahir Basharat Cheema, Tahir Saleem, past President and Vice president respectively of IEEEP who gave me the opportunity to speak and test my hypotheses and propositions in their seminars.

I am grateful to Saadia Qayyum who painstakingly edited the manuscript without her assistance, this book could not have been in the form it is now. I cannot continue to live at my home, if I do not mention my wife Dr. Meher and my beautiful daughter Schanze' who provided various kind of assistance in data searching, photographing graphs and tables and other work. My wife Dr. Meher has been editing my earlier works, but this time she has reviewed the book only advising me to soften statements which she thought were rather harsh and could be better stated alternatively. My wife deserves special thanks for almost being pushed out of the living room which I occupied for all the period in which I wrote this book. I am grateful to her for showing understanding. However, there still may be some sharp statements or criticism left, for which I would like the recipients to take it lightly. There is nothing personal about it. Finally, I should be held responsible for all the shortcomings of the book, and not to those whose names have been mentioned here or elsewhere in the references.

Syed Akhtar Ali
Islamabad
P | +92 345 2447714
E | Akhtarali1949@gmail.com
16[Th] July, 2017

List of Acronyms

AEDB	=	Alternate Energy Development Board
BCF	=	Billion Cubic Feet
COGE	=	Cost of Generating Electricity
CPPA	=	Central Power Purchase Authority
DISCOs	=	Distribution Companies
DOE	=	Department of Energy (USA)
EEX	=	European Energy Exchange
EIA	=	Energy Information Administration (USA)
EU	=	European Union
GENCOs	=	Generating Company's
GJ	=	Giga Joule
GOP	=	Government of Pakistan
GW	=	Giga watt = 1billion Watt = 1000MW
GWh	=	Giga watt hour = 1000 MWh
IAEA	=	International Atomic Energy Agency
IEA	=	International Energy Agency
IRR	=	Internal Rate of Return
KESC	=	Karachi Electric Supply Corporation
LCOE	=	Levellised Cost of Electricity
LESCO	=	Lahore Electric Company
LNG	=	Liquefied Natural Gas

LPG	=	Liquid Petroleum Gas
MCF	=	Thousand cubic feet
MMBtu	=	Million British thermal unit
MMCFTD	=	million cubic feet per day
MPNR	=	Ministry of Petroleum & Natural Resources
MTOE	=	Million Tons of Oil Equivalent
Mtpa	=	million tons per annum
Mtpd	=	million tons per day
MW	=	1 million watt = 1000kw
MWh	=	Megawatt hours = 1000 kWh
NEPRA	=	National Electric Power Regulatory Authority
NTDC	=	National Transmission & Dispatch Company
OGDCL	=	Oil & Gas Development Gas Company Limited
OGRA	=	Oil & Gas Regulatory Authority
OIL	=	Oil India Limited
p.a.	=	per annum (year)
PAEC	=	Pakistan Atomic Energy Commission
PEPCO	=	Pakistan Electric Power Company
PML	=	Pakistan Muslim League
PPIB	=	Private Power & Infrastructure Board
PPP	=	Pakistan People's Party
QESCO	=	Quetta Electric Company
ROA	=	Return on Assets
ROE	=	Return on Equity
RT	=	Refrigeration Ton = 8000 Btu/hr

SNGPL	=	Sui Northern Gas Pipelines Limited
SSGC	=	Sui Southern Gas Company
TAPI	=	Turkmenistan, Afghanistan, Pakistan, India (Pipe Line)
TCF	=	Trillion Cubic Feet
TOE	=	Tons of Oil Equipment
WAPDA	=	Water & Power Development Authority
WCA	=	World Coal Association

PART I: GENERAL

1. Introduction

There are three major developments in the energy sector, which warrant a fresh appraisal of policies and projects and thus the aim of this book. These are;

- Advent of CPEC
- Drastic oil and gas price reduction
- Emergence of renewable energy as a viable energy source

New issues are on the table, some have sustained over the years and some of the older issues are no more. Pakistan is marching towards success in solving its energy crisis. 10,000 MW of electricity projects are at various stages of implementation and are going to be commissioned in 2018. Similarly, several RLNG and pipeline projects and Thar coal development are also at advanced stages and still more are coming in. PML (N) may lose elections, if projects are not commissioned early enough, but Pakistan would still succeed, as after all the required capacity would be installed, even if slightly delayed. It may be noted that these all are not CPEC projects, as there are quite a few which are not part of CPEC, although the initiative has played a significant role and is poised to deliver even more in coming days.

If the objectives of CPEC were achieved, it would be the second marvelous achievement in Pakistan's history. The first one was Pakistan's nuclear program, which was founded by PPP's Z.A. Bhutto, sustained by Gen. Zia-ul-Haq, and was concluded by P.M. Nawaz Sharif who finally conducted the nuclear explosion. He didn't shy away despite pressure, as did all others. It was a demonstration that Pakistanis are capable of mustering unity of purpose when it is required. Similarly, consensus is emerging and evolving on CPEC even though there are understandable problems regarding distribution of development effort. CPEC is a long-term project in which all parties and shades of opinion will have a chance to play their role and make contributions. PPP as well as Gen. Musharraf have claimed that CPEC or Pre-CPEC was conceived in their period, but the fact remains that PML (N) put its act together on a fast-track and finalized a frame-work and managed to initiate many projects without wasting time in preliminaries which have resulted in problems. There are pros and cons.

There is some criticism or concern that Pakistan's debt profile is moving towards a critical zone. This may be a rather quiet criticism on CPEC and other development projects. The issue is that you cannot launch development projects of capital-intensive nature without external borrowing. The same economists who have made debt calculations had also made calculations that Pakistan's GDP has suffered a reduction in growth rate of 2% due to energy short falls. They should also see, what would be the impact of road connectivity on trade, employment and growth and develop a composite and unified picture, which may indicate where to stop or modify. Scaring the people by only adding negatives may not be expected from sincere economists who have in the past benefitted from opportunities in countries social and economic life.

On the other hand, the role of opposition parties is a must in keeping the government and functionaries on track. There is a lot of vested interest that cannot be controlled by well-meaning people in a ruling party, the scare and threat has to come from the opposition, civil society and the press. That is how a democratic system works in every society, which has managed to develop under a framework of freedom and liberty.

Present government had two objectives in its energy policy; to reduce load-shedding by increasing the generation and allied capacities; and to lower the cost and prices by reducing reliance on expensive Furnace Oil, which they did by introducing Coal power and RLNG. The latter became cheaper in mid-course and then they opted to install super-efficient NGCC plants with thermal efficiencies of 53%, unprecedented up till now. In order to have fast track implementation, they financed it out of PSDP avoiding any negotiation with external agencies. By far, the present government succeeded in its strategy. Chinese cooperated as well by fast track implementation of the coal plants. Imported coal and local Thar coal projects (3 in number) are being implemented. And even more projects on Thar coal are at various stages of development. It cannot be denied that success is on its way in this respect. We will discuss the details in forthcoming chapters.

The Challenges in Energy Sector and how to meet those

Tariff Issues: While it is highly commendable that energy crisis situation is about to end with the completion of many projects that are under implementation, there are challenges that remain and would have to be sorted out. Firstly, with the increase in supplies, subsidy and circular debt may increase, unless theft is brought under control and hopefully generation cost is brought down with the advent of new projects. NEPRA had to play a major role in containing generation tariff in which it has shown only a mixed success. In some cases (HVDC), it has succeeded in putting its foot down and resisted undue demands of higher tariff and in case of renewables, it has finally issued its edict on compulsory competitive bidding (for which I have been fighting for years by now) against the wishes of vested interest. In case of RLNG power plants, the tariff is almost right, but that is as a result of competitive bidding under GOP financing. In other cases, it has been swayed by vendor supplied data and could not undertake rigorous scrutiny. The case in point is of coal tariff, both of Thar and of imported coal power plants. While environmental issues pertain to more than one agency, it is hoped that NEPRA would take a corrective step and issue a new tariff which is due and apply the new tariff on all those projects that have not yet been implemented.

Many people believe that Coal Power tariff has not been determined appropriately and mishandled by the stakeholders on government side. Even if competitive bidding option is not available under G-to-G bilateral projects, it is still possible to reach agreement on reasonable terms and prices. It should be made sure that vendor does not influence policy. There are many ways and sources to find out the real price.

Two arguments are given defending the terms and conditions of the financial framework of CPEC, which may be partly correct; fast tracking requirements and that beggars cannot be the choosers. Fast tracking, it may be accepted that it precluded any long and stretched negotiations, however, many projects having been implemented by now, should give an opportunity to negotiate better terms. On the beggar argument, it can be argued that even beggars are able to get better deal from givers, if he or she presents his case appropriately. We know this from our normal life experience. And even beggars argue that alms giving is mutually beneficial as it relieves the alms-giver of guilt and provides relief and good feelings. In business also, it is not a one-way traffic. Interests and benefits are mutual, although one party may have more options than the others. There is now time to negotiate better terms on energy projects, especially on coal, as has been indicated in relevant sections. If CAPEX are high due to no Competition, then financing terms have to be soft as is G-to-G contracts. You cannot have commercial financing terms and as well as the luxury of no-competition G-to-G cozy environment. One can and should be able to put it across in a nice and pleasant manner without breaking the deal. Chinese companies would understand. They know the real prices themselves. In some cases they are getting prices almost twice as they get at home. Planning Commission of Pakistan and NDRC of China should be able to develop and negotiate a framework that is fair and sustainable.

Bhasha Dam: The most important item that should be on our national agenda is the construction of Bhasha dam. KalaBagh dam could not be built due to internal political differences and Bhasha dam could not be built because of external factors. We should go on our own. Ethiopia has built a very large dam out of its own internal resources without resorting to borrowings from IFIs. . Ethiopia faced almost the same problem as Pakistan is facing to construct GERD.IFIs and others wouldn't finance it due to potential and real objections and concerns of its big neighbor, as IFIs require NOC of India as a precondition for financing.

Ethiopia had no option but to rely on its own resources. It floated internal bonds and made deduction from salaries as loans and financed a 6.43 Billion USD project. It started construction in 1911 and is about to be completed and commissioned. We cannot do exactly what they did .But we need not. Our national and private banks have become quite big. Instead of financing sundry projects, they can be made to finance Bhasha. In fact, they have offered to do it to WAPDA which has proposed a phased strategy for going local. Chinese have offered to build it too. But they have proposed a very ambitious proposal to buy-out the whole Indus cascade which may attract a lot of concerns and in turn can cause delays. Let us go alone in financing. Eventually, others would join. It is hoped that GOP proceeds as promised that it would start construction next year which is the current year. However, in order to build Bhasha Dam, hasty decision on awarding the full Indus cascade to Chinese should be avoided.

The New Order: Solar and Wind Power: Solar and Wind Power have tremendously improved their competitiveness; their cost of generation have almost become 50% of the fossil energy based power. In Pakistan, their induction has been partly obstructed by unreasonable demands of the investor lobby. Hopefully, competitive bidding will break that circular situation and the Solar and Wind power would be available at its true cost and prices. A fresh thinking is to be given for a large scale induction of these resources to be able to bring down the cost of generation. There is resentment in the solar and wind power circles that while competitive bidding has been announced, no practical steps are being taken towards arranging it. The renewable energy investors also complain that RLNG plants are being approved one after the other, and the recent one has been done for Jhang of a capacity of 1100 MW, while renewable energy projects are being put on a back burner under the excuse of impending capacity glut or trap. Since many of such projects are to be located in Sindh, the issue assumes a political dimension and creates tension within the federation.

Solar Plan: Most energy and electricity planning has focussed on demand, supply and economics. Logistics and spatial planning has been ignored. Distributed generation requirements of Solar and water requirements of fossil power plants should be factored in a Spatial Plan which allocates power generation capacities and water withdrawal quotas. Solar Power should not be generated in the manner and style of fossil power. It should be generated in a distributed manner.QA Solar Park may have been a good beginning. However, distributed solar generation close to population clusters should be preferred for which 30-50 locations should be identified in the proposed spatial plan.

There is a strong case for planning a package of 5000 MW for 50 sites. Such studies should involve GIS technologies .Admittedly, this plan cannot be prepared in isolation. It should be associated with load studies. As a result, long hauls in Electricity transmission should be and would be discouraged saving energy and financial resources. Reportedly, HVDC project has been shelved due to good reasons, although, for this some coal projects had been introduced in Punjab in order to be able to obviate long haul transmission. Interestingly, coal transmission costs are almost the same as electricity transmission under HVDC.

Indigenization: indigenization has suffered due to the fast tracking dictates of the energy crisis. Now that the crisis is about to end, some stress and emphasis should revert to indigenization. Local engineering industry is heavily underutilized and its human resource base is contracting due to low market demand. Incentives to local industry and disincentives to imports in the form tariff and non-tariff barriers would have to be created. Chinese have played a major role in building our engineering industry in days when they were not commercialized. However, they can still be persuaded to rebuild our local engineering industry when tremendous demand is building up in the sector. JVs and buy-outs in engineering sector may have to be promoted. Energy tariffs and incentives may have to be tied together and deletion programmes put in place ala automotive industry, for which some standardization of energy equipment may have to be undertaken. In the current style of things, we are behaving like our Arab brothers, who place orders and foreigners provide it all from EPC and O&M contractors to IPPs. Iran and Turkey have done otherwise and are reaping the benefits of their policies. They are increasingly getting active in export markets including Pakistan, although Iran has other problems due to which it cannot sell abroad.

Local Resource Development: Energy crisis and thus the fast tracking requirements have resulted in temporary de-emphasis on local resource development. The most glaring case is of an almost imminent diminution of known gas reserves. Adequate resources and attention should now be given on E&P of Gas, if not Oil. There are abundant gas resources that are available to be explored and exploited. Consideration should be given to forming additional companies for this purpose in JVs or otherwise. Good financial status and performance of OGDC and PPL is not enough. They should be asked to raise their activity level. Building of energy future on Gas ala other countries is not a good option. There should be a moratorium on further RLNG Plants beyond the already planned or approved projects. Gas prices, along with Oil, will acquire their previous high level in a decade.

Building Export Projects along with energy import projects: Every action has both positive and negative consequences. A matter of concern is the projected debt servicing liabilities and even imported fuel bill. This would certainly create a lot of strain on

Pakistan's balance-of payment capabilities, unless exports are raised concomitantly. Normal export increase prospects do not appear to be very high. It may be a good idea to create export projects meant for export to China. One of such possibilities is export of minerals, particularly, copper. Readers may recall the Reko Diq mining project, which became victim of the immaturity of a motley crowd of communists, nationalists, ill-informed armchair economists and above all an activist judiciary of the time. Some other enthusiasts raised unrealistic optimism for self-development and revival of the Reko Diq project, little realizing that it requires several billion US dollars of investment, technology, management skills and marketing network.

Reorganization of the sector: Power sector, if not the oil and gas sector, is fast losing technology absorption and accumulation capability due to lack of appropriate organizational structure under the garb and slogan of de-bureaucratization. PEPCO has been dismembered and the sector is being micromanaged from Islamabad by non-technical people. Power sector has to be reorganized with a technology orientation: either MoWP has to induct technical resource persons under an appropriate organizational structure or separate organizations have to be created. Utopia of independent boards and NEPRA tutelage has not delivered. NEPRA can only issue edicts and orders and may be fine but cannot improve organizations and control them.

The Energy Task Force: There are too many issues requiring deliberation and consensus from all stake-holders. Energy issues are being dealt in cocoons of individual ministries and departments. It has to be seen in an integrated manner. Gas, Oil, Coal and Electricity affect each other. I have focused here on coal only, although this book attempts to touch all aspects. An Energy task force should be established to make recommendations on all the above-mentioned issues and other ones not discussed in this space. It should have representation from the provinces, bureaucracy and as well as professionals; federal and provincial energy ministers and some educated politicians should be included. This task force should be assisted by a secretariat composed of consultants and sector professionals. The issue of environmental controls is of immediate nature and should be attended without waiting for the task-force which may or may not be formed. In the following, we introduce individual chapters.

Strategic Directions in the Energy Sector

Pakistan has many energy resources and is in neighborhood of countries, which are among the richest in Energy. Distance plays a role. There are hydel and coal resources, which have mutually balancing characteristics. Sun always shines here with generosity and wind blows intensely in its season in a number of geographical areas. But no resource in itself is enough; it has to be developed though money, technology, skills and effort. It is not easy. Things do not occur either automatically or smoothly. There are vested interests and shear ignorance and negativity. There are also genuine problems, issues and difficulties, which have to be sorted out as one moves. This book is about all this. While it elaborates on achievements, it attempts to point out problems and issues and attempts to recommend alternatives and solution. The contents of the book are introduced and summarized below.

In Chapter 2: Strategic Directions in Energy Sector, we have painted a broad picture of the next 20-25 years of the energy sector including oil, gas, coal and electricity. We have tried to speculate on as to how demand, supply and prices would play out so as to draw some policy conclusions and recommendations. In order to be able to do that, we have provided past data as a reference point on which some plausible assumptions could have been made regarding the future growth and performance of the sector.

What is the Right Energy Source?

There is no right energy source: all energy resources have one limitation or the other. Solar is only in the day, Hydro and wind are available only in summer, Coal has environmental issue and there are other limitations in Thar limiting production possibilities to 10,000 MW, Furnace Oil is expensive and Gas resources are depleting and imported LNG is expensive and will cause foreign exchange drain. Nuclear has a large number of problems and limitations. All cannot be rejected on negative grounds, and all cannot be inducted 100 %. There is to be an optimal mix.

Concluding, as we do in Chapter 3: What is the Right Source of Energy, there is no panacea. All energy sources have positives and negatives. A composite optimal mix with the objectives of least investment and cost of generation has to be worked out and implemented with adjustments. Optimum does not always work. Supplier and investors preferences also matter in a country, which has not been able to attract FDI for a long time. Local resources are to be preferred even if costing a little more to avoid foreign exchange drain, but should not be a license to promoters and investors to indulge in undue profiteering and capital padding. Coal and other fossil resources are on the way out and Solar is the future. The future, though, has not arrived yet. Let us slide into the future gradually and not jump into it to get hurt.

Energy Access in Pakistan

Access to clean, reliable and modern sources of energy is critical for achieving the targeted economic growth and development in a country. Vision 2025, formulated by the Ministry of Planning, Development and Reforms of Pakistan, recognized the importance of energy access; Pillar IV of Vision 2025 aims to ensure uninterrupted access to affordable and clean energy for all sections of the country's population.

Pakistan is yet to achieve its access goal, as the national average electrification rate is 68%. More than 80% are in rural areas, where a World Bank survey found that 30 to 45% of households use kerosene as a primary or secondary source of lighting. According to the National Electric Power Regulatory Authority (NEPRA) State of Industry Report 2016, more than 32,000 villages in the country continue to remain without access to the electricity grid, forcing the residents to use traditional sources of energy, including firewood, kerosene and diesel, for meeting their lighting, heating and cooking needs. In areas such as these, it would make economic and financial sense to set up decentralized plants near the load centers directly supplying electricity to the consumers.

In Chapter 4: Achieving Universal Energy Access in Pakistan, we present national and provincial level statistics on energy access as well as provide recommendations on how we can achieve universal energy access by transitioning to modern fuels.

Are Energy prices high in Pakistan?

This is a very complicated, often-asked and yet an extremely important issue. An all-embracing global answer cannot be given at this stage of data and indices availability. However, limited answers can be given and comparisons can be made fuel wise with individual countries. Taxation component often makes the country comparison difficult. Some countries apply high taxes, especially, on petroleum products, and some do not. In the U.S., energy is taxed minimally and their prices are the lowest. In India petrol prices are much higher than in Pakistan, but diesel is lightly taxed with resulting lower prices. In In Chapter 5: Are Energy prices High in Pakistan?, we will attempt to make some specific comparisons in order to be able to deliver definite answer and try to explain and explore, as to why and where are the differences, if any.

Energy Prices and Economics in Pakistan

There is a wide difference in costs and prices of various energy sources; coal vs oil vs gas etc. There are factors other than taxation which are responsible for these variations. We will examine in Chapter 6: Energy Prices and Economics in Pakistan, the cost structure of various energy sources that would enable us to develop an understanding and appreciation of the differences involved. Competitive factors cause a difference among cost and prices. We have examined this issue in another chapter e.g., the case of oil where large margins have enabled profits even when sales price has gone down by more than 50%.This means monopolistic influences on oil market which collapsed with contraction of demand. There is a controversy on Thar Coal and Energy tariff, the latter we have compared with imported coal tariff and that prevailing in India on Lignite and others. We have also provided comparison between RLNG and Coal based energy costs and tariff.

Where will Oil and Energy Prices go?

In Chapter 7: Where Will Oil and Energy Prices go? , we examine yet another important issue of the future oil price behavior, for it will affect almost everything. Oil and energy in general play a major role in our lives and economies today. People cooked their food on wood and shrubs, as they still do in most of our rural areas even today. The tradable energy is what is new and that is why the subject of energy prices is so important.

Oil's dominance can be measured by the fact that seven of the largest ten multinationals are oil companies, with Royal Dutch Shell being the largest at 275 Billion USD in sales and revenue. The two super powers are among the largest oil producers, and one (the U.S.), being the largest importer and consumer of oil consuming one-fifth of the oil production.

If you are a consumer or an importer, you would like that oil and energy prices are as low as these days and continue to do so for as long as possible. On the other hand, if you are a producer or an exporter, you would like the reverse to happen. In Pakistan, purchasing power is so low that even low prices are not affordable. But the governments want to extract as much from oil as it can. We would like to apprise the reader of the possible oil and energy prices scenario in a context of the next two decades and the behavior of factors that go into. A tall order for the space that is available.

Priorities in CPEC

CPEC is playing a very important role in Pakistan's energy and infrastructure sector and hopefully would continue to do so in times to come. New projects are being approved on roads, circular railways and industrial estates. In addition to the energy sector, every province, with the help of China, will establish one industrial estate, and one rail-based mass transit scheme. There is also a long list of miscellaneous provincial schemes, which many have found slightly out-of-place in a larger framework such as of CPEC. It was a good omen, however, that Chinese have not been discouraged by the experience hitherto and have come with more. In Chapter 8: Priorities in CPEC, we draw attention to issues that merit some consideration for prioritization in the CPEC framework.

Financiers always give importance to the returns of the project, although recipients are not usually very sensitive on this as for them it is "Qarz ki Mai" (wine on credit as per ghazal of Mirza Ghalib). Provinces should not insist on unviable projects and should give importance to the financial viability. After all, it is they who would have to pay, unlike earlier CPEC projects of federal content and nature.

Every action has both positive and negative consequences. A matter of concern is the projected debt servicing liabilities and even imported fuel bill. This would certainly create a lot of strain on Pakistan's balance-of payment capabilities, unless exports are raised concomitantly. Normal export increase prospects do not appear to be very high. It may be a good idea to create export projects meant for export to China. One of such possibilities is export of minerals, particularly, copper. Readers may recall the Reko Diq mining project, which became victim of the immaturity of a motley crowd of communists, nationalists, ill-informed armchair economists and above all an activist judiciary of the time. Some other enthusiasts raised unrealistic optimism for self-development and revival of the Reko Diq project, little realizing that it requires several billion US dollars of investment, technology, management skills and marketing network.

The Conundrum of Single Bid

Over the last few years, we have seen many projects being implemented under single bid approach, which in some cases has resulted in rather excessive costs. However, if the funding is available under a bilateral arrangement, there is possibly no escape from single bid projects. Competition can be organized if funding is independent of supply, which is only possible when the funding is from third sources or multilateral institutions.

IFIs or multilateral institutions are not the best choice either as they do not have the kind of resources that have been made available under CPEC. There is a long processing cycle in IFIs and the funding is associated with conditionality and agenda that may be compatible with international frameworks, but often in terms of scope and timings, not realistic or compatible with the day to day running requirements of the country. Moreover, they always require some kind of surgery. Irrespective of these peculiarities, as we have mentioned, IFIs have to loan to many borrowers and thus cannot cater to the required level of funding.

The issue is not just limited to CPEC; it is common to all bilateral funding such as from Russia, UAE, etc. in the past. EXIM Bank and the other US banks partially have attempted to tackle this issue, where in they require competition among the bidders, although restricted to the American suppliers. Perhaps Chinese can be prevailed to do the same. They have a large economy and many suppliers of single products. They are encouraging and enforcing competition in the domestic sector. It may not be a bad idea that China tries the same in CPEC. We have explored the possible options and strategies to deal with this difficult issue in Chapter 9: The Conundrum of Single Bid.

Are We Consuming Energy Efficiently?

Due to depleting fossil fuel resources in the world and the climatic effects of burning these resources, there is a lot of stress these days, on countries, to reduce energy consumption. In developing countries like ours, who have only begun on the road of industrialization, the only option is to be in harmony with the world community is to improve our energy efficiency by controlling wastes and producing and consuming less CO_2 producing electricity. Thus energy efficiency is to be better understood by all of us to be able to better contribute towards achievement of this objective. In earlier days of development and industrialization in the world, energy consumption was considered an index of development. It still is in many ways. United States, China, Russia are the largest energy consumers. In Chapter 10: Are We Consuming Energy Efficiently? We examine the options and strategies in this respect.

Towards a Realistic Climate Change Policy

Climate Change is a catastrophe that can be avoided by combined human action and cooperation, as human civilization has been able to tackle many issues that threatened its well-being and survival in the past. However, what level of effort and sacrifice has to be made by individual countries has remained a matter of contention.

Climate Change is caused mainly by the accumulation of Green House Gases (GHG) such as $CO2$, SOx, NOx, and Methane etc. Predominantly, it is $CO2$, which is a product of combustion of fuels and biodegradation of organic materials. These emissions lead to rise in temperature of earth and atmosphere affecting metrological balance and melting of glaciers etc. causing droughts, floods, hot summers, and rise in the ocean level threatening inundation of many coastal cities like Karachi and others.

Developing countries including Pakistan have been in confusion as to what should be their target and what kind of commitments can be made and even estimating the emission levels and possible reduction without damaging their economies has been beyond most developing countries. India submitted its INDC last year and Pakistan could do it only last month. The issue still remains as to whether any reduction in emissions can be targeted, as our current emissions or even projected ones say of 2035 are going to be too low. Reduction in emissions say from 2035 level may tantamount to limiting our development. Pakistan's Climate Change policy makers, perhaps under pressure of announcing some commitment and to be the part of international process and to be able to benefit from Mitigation assistance packages that are to ensue, have submitted INDCs, committing to a reduction of up to 20% of its 2030 projected GHG emissions subject to availability of funds.

In our view more creativity and effort should have gone into estimating and deciding on INDC commitments. We could have reviewed INDC of other countries, especially, of the region, particularly India. India has not agreed to any absolute limitation on its emissions at all. but has only agreed to reducing its emissions per unit of GDP by 30% by the year 2030.This is reasonable as this amounts to improving energy and emission efficiency by implementing conservation policies. Pakistan should also make its INDC submissions on these lines. Things are not at a level of finality at this stage. There is still time to come up with emission reduction targets and approach that is consistent with our unrestricted rights to growth. We can show our sincere commitment to the international community by agreeing to promoting conservation, renewable energy induction and development of emission sinks like forestation. This would be taken more seriously than a kind of commitment that eventually may not be implementable. The reader would find more details in Chapter 11: Towards a Realistic Climate Policy.

Competition in Renewable Energy

Utilities were long considered as natural monopolies as these used to be integrated companies combining generation/supply, transmission and distribution. These days, integrated utilities are rare and generation of electricity and supply of gas is universally competitive. Distribution companies buy electricity at competitive rates largely determined in energy/ commodity exchanges in a stock market like situation. In developing countries, however, there are regulatory agencies like NEPRA/OGRA, which determine the prices that are fair to both producers and consumers. Many people in the country are skeptic of competition as has been indicated by frequent raids of even a docile Competition Commission of Pakistan. This has created justification for continuation of regulation.

High solar and wind tariff have largely been responsible for impeding the growth of the renewable sector, in particular, the solar and wind. There was and is no need that solar and wind tariff be kept so high as compared to other countries under vendor influence, flawed regulatory practices and reliance on public hearings without making use of good third-party expert advice. Although regulation is almost always less efficient than competition, regulation could have been better conducted. The above should not be construed as indictment of the present leadership either in government or the regulatory agency in question. It has been a common problem over the years. Let us look forward to better days and an appropriate tariff under competitive conditions.

Time has come to introduce some competition as we will see in the following the failure of regulation. Despite one's reservation against competition, which can be manipulated and hijacked, the world experience is that competition drives down prices more than regulation. Prices are discovered through bidding. Otherwise, there is no more accurate way of knowing the prices of inputs and outputs. Cost estimates by regulators or bureaucrats may be faulty, in favour of supplier or the buyer. It augurs well that NEPRA, after observing all that has happened in the past and facing difficulties in arriving at the right tariff, has come out in support of competition, although a bit late as they say, *dair ayad, durust ayad*. NEPRA has issued guidelines for competitive tariff for solar and wind projects. In Chapter 12: Competition in Renewable Energy, we have examined the allied issues and have made recommendations.

Bhasha Dam: Beg, borrow or steal

The most important item that should be on our national agenda is the construction of Bhasha dam. KalaBagh dam could not be built due to internal political differences and Bhasha dam could not be built because of external factors. We should go on our own.

Ethiopia has built a very large dam out of its own internal resources without resorting to borrowings from IFIs. The dam will be able to store 79 km3 of water and produce 6000 MW of electricity. For comparison sake, Tarbela had a storage of 13.69 KM3 which has come down to 7.993 km3 due to silting. The proposed Bhasha and KalaBagh dams are almost equal with 7.52-7.9 km3.Infact, as would be evident from the adjoining table, even if we build all the dams proposed, it would total to 42.3 km3 which would be only half of this dam.

Ethiopia is much poorer and smaller than Pakistan having a GDP 25% of Pakistan. It has suffered famines and civil wars. I have visited Ethiopia several times doing my professional consulting services for building parts of the power house locally in Ethiopia. River Nile springs from its Lake Tana which flows through Egypt and ten other countries including Sudan and falls into the Mediterranean Sea. The erstwhile General Nasser built ASWAN dam on it. Egypt is largely dependent on Nile water and is worried over the consequence of the GERD dam, with a storage capacity as large that it may require 15years to fill to its full. Ethiopia faced almost the same problem as Pakistan is facing to construct GERD.IFIs and others wouldn't finance it due to potential and real objections and concerns of its big neighbor, as IFIs require NOC of India as a precondition for financing.

Ethiopia had no option but to rely on its own resources. It floated internal bonds and made deduction from salaries as loans and financed a 6.43 Billion USD project. It started construction in 1911 and is about to be completed and commissioned. We cannot do exactly what they did .But we need not. Our national and private banks have become quite big. Instead of financing sundry projects, they can be made to finance Bhasha. In fact, they have offered to do it to WAPDA which has proposed a phased strategy for going local. Chinese have offered to build it too. But they have proposed a very ambitious proposal to buy-out the whole Indus cascade which may attract a lot of concerns and in turn can cause delays. Let us go alone in financing. Eventually, others would join. It is hoped that GoP proceeds as promised that it would start construction next year which is the current year.

In <u>Chapter 13: Bhasha Dam : beg, borrow or steal</u>, we make a strong case for Bhasha dam in addition to presenting an expose' of the hydro sector including hydro power potential of Pakistan, status and implementation of various projects and the economics of hydro power in the backdrop of falling prices of Solar and Wind Power. It appears that Hydro Power is fast losing its classical competitiveness, from a Tarbela Tariff of Rs 1.0 or less per kWh to Rs. 8-10 per kWh, while Solar and Wind Power are today typically costing under or around Rs.4-5 per kWh. Also long gestation time of Hydro has always been an issue. The low prices of Solar and Wind are a relatively new phenomena. It is quite possible that the future decision makers may prefer going for Solar and Wind in preference to Hydro Power, except where water storage is involved.

Hydro Power Royalty

Hydro Royalty or Net Hydro Profit (NHP) as it is called in our constitutional (1973) terms, has been a major dividing and polarizing issue. Some provincial leaders of KPK have issued extremist and incendiary views on the subject demanding early resolution and implementation of their demand in this respect. CCI has reached a consensus on it and its decision has been passed on to the relevant authorities. Although consensus has been reached on numbers, which is to be welcome, it is temporary. There is a need for building a lasting and sustainable solution based on definite definition and methodology on which there should be consensus. It should not be too less for the recipient and too much for the payer and producer and should be comparable with other jurisdictions. Some political accommodation would have to be reached on constitutional niceties as well. It will be a tall order, but can be done. We have discussed the issues in detail in Chapter 14: Hydro Power Royalty.

Restructuring Oil and Gas Sector

Oil and Gas (O&G) sector has defied reforms and restructuring for a long time now, despite demand by the stakeholders and recommendations of others. Even 18th Amendment could not move them. It is a big joy lording over such a big and in many ways quite a prosperous sector. So much so that recently, some wise men of MPNR proposed investments and activities in power sector as well.

Over the years, the power sector has undergone major restructuring from a monolith of WAPDA to DISCOs and GENCOS. More reforms are in continuous deliberations. On the other hand, the O&G sector has not progressed as is indicated by deteriorating performance of the gas companies, depleting gas reserves, and underdeveloped transmission infrastructure. Exploration companies have been complaining all along of bureaucracy and red tape in petroleum concessions. Fortunately, Minister Khaqan Abbasi has turned his attention to it and is in favor of a fast track change. World Bank consultants have been appointed who are in the process of consultation and finalization of their recommendations. The purpose of Chapter 15: Restructuring Oil and Gas Sector is to inform the stakeholders of the issues involved and posit some humble recommendations.

Natural Gas and LNG Demand

Natural gas has emerged as attractive and popular energy source. Firstly, oil and gas prices have come down. Natural Gas has traditionally been sold at 75-80% of the oil price. It is expected that this ratio would sustain. Secondly, LNG prices have also come down and availability has increased. And thirdly, Combine Cycle Plants' efficiency has increased phenomenally up to almost 60%, almost twice or more the present level of average efficiency. Due to its relatively clean burning characteristics and lower carbon footprint, it is also being preferred by many countries.

We are passing through a general energy crisis, of both gas and electricity. Apparently, there should be no reason for joint occurring of both. Lack of electricity does not cause reduction in gas production, although reverse may be true. Major expansion has been launched in Pakistan. 5 LNG plants are in pipeline with a total expected capacity of 2-3 bcfd (400-500 mmcfd gasification capacity per plant).

Gas production has stagnated for almost a decade since 2008 at around 4 bcfd. It was preceded by a growth period of 2002-2007, in which gas production increased from 2.5 to 3.8 bcfd. Ironically, reverse has been the case of oil; in almost the same growth period (2002-11) of gas, oil production stagnated at around 65,000 barrels per day. Thereafter, Oil production started increasing rather phenomenally and reached a production level of around 100,000 barrels per day in just 5 years. Why production in one has caused stagnation in the other? I have heard, the exploration effort is common; it is almost a chance that either oil or gas is discovered. Oil is useful, even if there is no oil refinery in the country or nearby. It can be exported using trucks and ships. But gas is useless unless there is a gas processing plant nearby or a new one is installed. It takes around seven years to explore and develop a gas field, although oil may require less. Hence, it is important to develop a reasonable forecast. Unfortunately, I have not come across a reasonable study on the subject unlike electricity where several major studies have been done which we have referred to elsewhere in the electricity section. We will undertake here a rough-cut analysis in Chapter 16: Natural Gas LNG Demand.

Non-Conventional Gas Resources

There are following non-conventional gas resources that have potential of varying levels for filling the demand and supply gap;

1. Shale gas: 95 TCF
2. Tight gas: 24-40 TCF
3. Stranded gas: 500 MMCFD (3 TCF)
4. Flared gas: 150 MMCFD (3 TCF)
5. Coal bed Methane (CBM): 20 TCF
6. Existing Conventional resources: 25 TCF

There are environmental and climate implications to all these gas resources. CBM, if not utilized, can leak into the atmosphere. Methane has 35 times higher climate footprint than the most blamed CO2.Secondly, gas being climatically benign relatively, its higher availability in the energy mix would have impact on national climate footprint. Stranded, and Flared gases are immediately available resources that can be utilized by new technologies such as Mobile LNG/NGL production facilities. Shale Gases have a long term potential. There are issues of water requirements and disposal and of availability of technologies which are tied in a few hands internationally. We examine these issues in Chapter 17:Non-Conventional Gas Resources.

Alternative Approaches in RLNG

Five RLG plants have been planned. One has already been installed at Port Qasim. Contract for another has been awarded which is expected to come online within this year. Three more RLNG plants are in the pipeline, of which one at Gawadar is at an advanced stage of planning. Although there has been a meaningful progress and performance in RLNG, we would like to examine if there are alternative and better approaches in building further RLNG capacity.

Alternative financing and even technology options have been explored and successfully implemented by India and Singapore and others. Lithuania, being a small and more conscientious country, has been trying to convert its lease agreement replaced by cheaper debt finance. However, it has not been successful, due to the unwillingness of the lessors. It is difficult to convert or change financing mode during mid-course, but it is certainly feasible in the beginning of the project, as is the case of our new projects. If RLNG is to be a major cornerstone of our energy supplies as it is emerging, risks are to be evenly distributed. It may be a bit too risky to have all the RLNG projects on lease periods of around ten years. Some projects must also be on long-term ownership or IPP basis. On ownership-debt terms, one may have a tariff as low as 25 cents per MMBTU or even less, as opposed to 66 cents and 41 cents respectively for Engro and PGPL. In Chapter 18: Alternative Approaches in RLNG, we examine the allied issues in the afore-mentioned perspective.

Alternative Approaches to LPG-Air Mix Plants

MPNR intends to install 65 LPG Air-Mix plants, at a cost of Rs. 17.6 Billion, to supply LPG-Air Mix to consumers in areas where Natural Gas (NG) distribution network is not available. Under the scheme, LPG will be transported to such areas by LPG tankers already established in LPG supply chain in Pakistan. Liquid LPG will be gasified in LPG – Air-Mix plants and connected to the isolated distribution network installed for this purpose. LPG-Air-Mix would be supplied to consumers in designated areas at a tariff that is normally chargeable for NG supply. The relevant gas distribution company in its annual revenue requirements, to be approved by OGRA, would absorb the expenses.

Merit of the scheme is that far-off areas would be supplied with a viable fuel for normal household use, which otherwise would normally use wood-burning contributing to deforestation and both indoor and outdoor pollution. Quetta remained without NG for quite some time due to pipeline economics constraints, which contributed to national disharmony and political problems for national governments.

The scheme has been criticized among some circles on the following grounds: It may be quite expensive scheme both in CAPEX and OPEX terms and the commodity charges. LPG is as expensive as Petroleum products (11-12 USD per MMBtu). With additional expenses of LPG-Air-Mix Plants, it would become more expensive. Practically Gas companies would be buying LPG at 11-12 USD and sell at 3-4 USD per MMBtu. Won't it be better to sell RLNG of 6 USD and sell it at 3-4 USD per MMBtu? Even though other consumers, through a cross-subsidy tariff, are absorbing such costs, it does entail costs in economic terms. Cross-subsidy may be opposed by OGRA; legal issues may also emerge, as LPG and NG are two separate products and businesses.

In Chapter 19: Alternative Approaches to LPG-Air Mix Plants, we examine some better options such as of SSLNG (Small Scale LNG) whereby imported (or even locally produced) LNG, could be transported to the off-network areas in a liquid form and re-gasify at distribution/user site. We have also identified some economic opportunities for producing LNG at NGL plants, which have already been installed at various gas fields. We also have made submissions for converting CNG stations to LNG by adding LNG storage and regasification facilities to the existing or new LNG/CNG stations. In fact, there are possibilities that, to a large measure, CNG sector may become independent of GAS DISCOs by adopting SSLNG.

LPG subsidies for poor to avoid deforestation

Only 20% of households in Pakistan have access to a clean and convenient cooking fuel like piped natural gas. Another 1.5 million Households have access to LPG cylinders, which leaves most of the country unsupplied, except for the major urban areas. Thus the idea of supplying some kind of clean fuel in affordable prices should be received with sympathy and affection.

MPNR has launched a scheme of installing off-grid distribution networks for distributing LPG. Some 24 schemes are being planned. The idea and intention is good and is to be appreciated, however, the approach and methodology does not appear to be attractive and viable. In Chapter 20: LPG subsidies for the poor to avoid deforestation, we will examine the issues involved in the proposed scheme and would explore other options as to their feasibility. In another chapter, we have examined an alternative approach of distributing Truck-transported LNG to such off-grid networks, which is used in many countries. In this chapter, we have explored another option i.e. that of LPG cylinder scheme under a subsidy system.

LNG Controversy

LNG contract with Qatar has become a hot political topic. The contract was handled by the P.M. Abbasi when he was Minister of Petroleum although he continues to hold the charge of Ministry of Energy which has recently been formed to merge Ministry of Petroleum and Ministry of Power. The antagonists of the PML (n) and PM Abbasi argue that Pakistan and Qatar LNG contract prices are unduly high and there is a possible scope for corruption and rentier intermediary interest in the contract. In Chapter 21.LNG Controversy, we have examined and presented the facts and data to let the readers make their own judgement about the merits of these allegations.

I must clarify at the outset that there is no intention here to support or oppose any partisan view. The Conclusion is that, Pakistan-Qatar contract prices are comparable with those in comparable and relevant countries in South Asia. LNG prices are a complicated issue and not as simple and standard as Crude oil prices. There is no standard and uniform reference price as is there in Brent Crude accepted widely. LNG market is evolving and prices may come down. However, we would be bound by our long term agreements. Spot prices cannot be compared to the long term contracts, as these are variable seasonally, are not guaranteed and carry no risk on the part of the supplier and are thus bound to be lower than the long term contract prices. Long term agreements have their pros and cons as everything else has. The right approach appears to be buying both in Spot and as well as contract, optimising the unit cost. If price differences increase, there may always be a possibility of renegotiations within or outside the contract at political level. The foregoing still does not prove that somebody may or may not have made money in LNG deal. Third party facilitators, having long term service or sales contract, may always be able to earn some money from the sellers.

Curtailing Losses in Gas Sector

33

T&D losses in the gas sector have assumed alarming dimensions amounting to 15% of the gas handled by the two companies, namely; SSGC and SNGPL. The monetary value of these losses have been estimated at around 8-900 million USD calculated at an opportunity cost rate of 6 USD per MMBtu of LNG prevailing price. These losses can be reduced and savings to the tune of 500 million USD can be generated, if corrective measures are taken. In Chapter 22.Curtailing Losses in Gas Sector, we have considered and evaluated various options and means to control the Gas sector losses.

Reorganisation of the Gas sector has been on the table for a long time now. Many studies have been undertaken and yet another with the assistance of the World bank most recently along with a parallel study by KPMG. The consensus has been to have one or two transmission companies, and the distribution function be divested into smaller Gas-Discos ala power sector. Smaller geographical domain may enable the top management to have tighter and direct control of field operations. A common problem in both power and gas sector are inadequate control metering. Meters are to be there at all nodal points in order to be able to do control accounting. It is not enough to know the losses at company level; there has to be control metering for every small geographical segment ,say, of 1000 customers enabling the respective managers to know how much gas has been shipped and how much billing and receipts are there in their respective areas.

Replacing leaking pipes and old meters and installing meters at all control points such as TBS requires money. Cash starved companies cannot invest money and nobody lends to such failing companies. Hence, it may be advisable to provide a reasonable tariff and UFG allowance. However, it is a double edged sword. A liberal UFG may provide further incentives, opportunities and market for company employees to collude with gas thieves. Hence whatever be the UFG, it is to be accompanied by a programme of loss reduction and action plan that identifies activities that have to be done. And that is what consultants (KPMG) seem to have done in there study. They have recommended a two part UFG allowance; 5% fixed and 4.05% variable under an action plan spreading over a period of 5 years. In the year 2015, SSGC had been awarded a UFG of 8.6 % and SNGPL of 7.1%.The proposed action plan consists of 25 actions, 15 of which are technical requiring investments and 5 are related with operational activities required for direct theft detection and control. However, a few aspects are lacking in the consultant's study (although, they are restricted by the TOR and the resources provided), in my view, and can still be undertaken.

Promoting Oil & Gas Production

It would be unfair to say that nothing has been done to promote local oil and gas production. Some effort, more as inertia and momentum of the past, has been done. Over the past three years, 319 wells have been drilled with 91 new finds. As compared to the past, this is a good record and performance, perhaps the highest than ever before. But it is not enough compared to the demand challenge and supply potential that exists. It does not match with the progress in the power sector; 20,000 MW Installed total up till 2013 and 10,000 MW in the period after. An initiative much larger in scope and strategy is required. What is required to be done?

In <u>Chapter 23: Promoting Oil & Gas Production</u>, we examine the question. The solution lies in creating market and competition and resolving governance issues. Look at what wonders have been done and are being done in KPK by KPK Oil Company. It has come out of autonomy of action and creating a third actor, apart from selecting competent persons to man the organisation. Half of the 21 active drilling rigs are reportedly working in KPK by KPKOC. They are targeting 2 Billion CFT per day of gas production by 2025.This has happened by letting to create a third entity. Let there be more entities. For example, MPNR should actively consider creating one or two more companies ala PPL and OGDC. An Oil drilling contractor company and a JV in oil service area. Local Private sector investment by smaller companies may be promoted after all they earn and lose in stock market Sattha daily. In Pakistan, one strikes one discovery for every three wells, not a big risk. Unfortunately, service companies are leaving Pakistan, as is the recent example of Baker Huges indicates. It may be due to overall restructuring and also because country presence is not considered necessary by the service companies in the information age. It costs 15-20 million USD to drill and complete an oil or gas well in Pakistan, which is 3-4 times more than elsewhere on the average. It can be brought down by creating a market of supply chain in Oil construction and service industry. Well construction costs would come down by creating the infrastructure and the market. Rigs can be partly produced locally except the drilling bit and rotary equipment. Local supply of rigs and installation know-how can be promoted. By these measures, existing companies can double their drilling output and possibly double the discoveries.

Enforcing Environmental Fuel Standards for combatting Pollution & Smog

Pollution and Smog paralyses life for many days in the autumn every year in major parts of Pakistan. Although it is a multi-dimensional problem, motor fuels have a significant role in all of this. In <u>Chapter 24: Enforcing Environmental Fuel Standards for combatting Pollution and Smog</u>, we will examine what we can do through stablishing and enforcing environmental fuel standards in the country, for bulk of the pollution and thus the smog generated through burning one kind of fuel or the other.

The whole world is shifting to low sulphur fuels. In Pakistan, only lately, we have been able to partly switch to low sulphur fuel satisfying the requirements of Euro2 standard, which is a rather old standard limiting sulphur level to 500 ppm. Most of the world has shifted to or in the process of shifting to even lower sulphur as required per Euro5 or 6 limiting sulphur to only 10 ppm. Industrialised countries have a problem of converting their refineries into low sulphur mode. However, we import most of our fuel from abroad and should not face problem in such switching. Low sulphur fuel is available in international market. It is a bit expensive but compared to the health and environmental costs, such extras are much less. India has already adopted Euro4 (sulphur level 50 ppm) and is poised to switch to Euro5 or 6 by the year 2020.Thus our gasoline and diesel will be having 10 times more Sulfur and thus 10 times more sulphur emitting vehicles on our roads, even after the complete switch over to Euro2 standards. I would argue for adopting 50 ppm standard (Euro4), and not be victims of the existing old refineries which cannot produce low sulphur fuels. Either they should be asked to import low sulphur crude oil or their output should be relegated to smaller towns and rural areas. All major urban areas should be required to be switched to 50 ppm sulphur standard through distributing imported low sulphur fuels.

Secondly, in many parts of the world, ethanol mixed gasoline is being marketed. E10 gasoline contains 10% ethanol and E5 contains 5%.In the U.S., almost all cars have switched to E10,mixing mostly bioethanol made out of sugar cane. This has an added advantage of having a RON rating of 94 or 95. As reported in the press recently, manganese and other metals have been added to raise the RON levels.

Problems and Prospects of Coal Power

Sahiwal coal power plant (1300 MW) has been commissioned and is working satisfactorily. My salute to Chinese that they have built a power plant of this capacity in such a short time. Government of Punjab has also been involved in solving administrative and logistical problems and deserves appreciation. Another similar coal power plant is at an advanced stage of construction and should be commissioned in the next few months. In Thar, Engro's Coal power plant is slowly coming up as well.

It is probably a stage where we should undertake some critical examination of some of the issues that have come up as decision about new projects is to be made. These projects were done in haste and it was natural that some due processes were avoided and lesser time was available to deliberate. I will try in this space to raise these issues and make some submissions and recommendations, although some of these I have been raising even earlier. These are the issues of cost of generation and CAPEX, environmental concerns and role of EPAs, NEPRA proceedings, imported vs local coal, location of future plants. We have discussed these issues in Chapter 25: The Problems and Prospects of Coal.

New Thar Coal Tariff

Coal Power tariff generally and particularly of Thar Coal has attracted quite some controversy. There are some genuine issues such as that of High RoR (20% in case of Thar coal and 16% on imported coal) and higher allowed CAPEX - which is a general malaise in Pakistan's regulatory system. However, NEPRA helped start a new controversy by introducing a new term of RoE in which returns (although equivalent) showed higher numbers, which could only have been appreciated by technically and financially articulate persons. RoE numbers came out to be 29-33% (equivalent to IRR of 16-20%), which created a lot of undue concern and anxiety. Doubts still persist among popular or even informed circles.

I am not sure why NEPRA helped start a controversy on coal tariff by approving RoE instead of IRR on Equity. I assume they are equivalent. In ROE, returns are provided only in operational years, while IRR includes payments during construction. I would suggest return to IRR which is more versatile, understandable and comparable and perhaps even more transparent. Secondly, tariff determinations are a bit cryptic i.e. calculations are not easily verifiable. Presentation should improve providing major calculation steps. One has to develop his own model in order to fully comprehend the determination issued by NEPRA.

People used to yearn for local coal utilization expecting that it would be cheaper to do that and would save foreign exchange but this hasn't proven true for Pakistan. Thanks to the add-on costs awarded by NEPRA to imported coal power tariff such as about 20 USD per ton for port handling, there is some comparability to Thar coal prices vs imported one.

Proposals for reducing Thar coal tariff should not be understood as being against Thar. On the contrary, a balanced approach is being argued. High prices increase producer's incentive. Low prices means buyer's incentive. Solar and wind promoters in this country kept lobbying for higher prices and got them, but resultantly, very few projects could be installed. History of other countries has indicated that market and investment expand at low prices. Thar promoters have to adopt a market sense. There is no monopoly of any energy source. If there was any, it has been broken by the imported coal which one would like to discourage, which compares quite unfavorably with Thar Coal power tariff of Rs.8.5 per kWh. Imported coal tariff should have come down automatically with reduction in international coal prices by 25-50%. Thar coal should be kept competitive with imported coal. Circular debt can only be eliminated, if the gap between cost of production and consumer tariff is reduced along with eliminating theft. All will have to participate proportionally in this effort. In <u>Chapter 26: New Thar Coal Tariff</u>, we make some allied submissions.

Environment and Coal Power

The irony is that when we have started exploiting coal, international opinion is going against it, but not without justification, as coal has adverse impact on climate change. However, our national contribution, even at the height of our projected coal production and electricity generation, may be very little as compared to others, say, India, China or even the U.S.A (the latter being more important in the wake of Trump Administration's anti-environment policies, which ironically may find quite some support in this part of the world, among some circles).

However, Health and Environmental consequences of large-scale coal power production should not be ignored. Four main emissions, namely particulate matters, SOx, NOx and mercury compounds, are released into the air by burning coal. Bronchitis, eye diseases, lung problems, heart attacks, mental disorders and stunted growth of children are some of the health problems that have been linked to these emissions. Moreover, NOx and SOx are washed down due to rain causing acid rains, which affect agricultural productivity. Mercury pollutes water bodies and finds its way into human beings through food chain. Pakistan has signed Minimata Convention in 2013, which aims at controlling or even almost phasing out mercury, as mercury pollution spreads far beyond national boundaries.

Responsible action can minimize these adverse health and environmental impacts. Pollutants can be arrested at source by installing appropriate equipment. High efficiency can reduce CO_2 production, which causes greenhouse effect. In Chapter 27: Environment and Coal Power, we deal with the allied issues and take to task the stakeholders in the environment sector who have been derelict in performing their duties, quietly allowing violations and not doing any whistle-blowing. We have found that it is the consultants who have not been performing their professional responsibilities adequately and coming out with justifications for avoiding installation of the required equipment and controls.

Underground Coal Gasification

There is an R&D project under which Thar Coal is being gasified underground (hence the name Underground Coal Gasification-UCG) for electricity production and possibly for producing other products such as fertilizers and chemicals. The project has managed to attract quite some attention as well as controversy. There are proponents and opponents having strong and weak arguments defending their positions. We will take stock of the project and try to guide the readers as to the merits and demerits of the arguments and the project itself. We will try to explain as to what the project is; what has been achieved and what has not been achieved and cannot possibly achieved; what is the international status of the project and what are the prospects of commercial interest of IPPs and developers to adopt UCG as a process and where does our R^D project stands in this respect; what are the options to benefit from the R&D efforts that have been done up-to-now. We have tried to explore answers to some of the questions and issues in Chapter 28: Underground Coal Gasification.

Industrial and Agricultural Uses of Thar Coal

In Chapter 29: Industrial and Agricultural uses of Thar Coal, we have discussed uses of Lignite other than large scale electricity generation which we have focused upon mostly throughout this book. There are many other uses of Lignite such as industrial and space heating, production of humic acid, soil conditioners and in making additives used in Oil and Gas drilling. Lignite has been used in industries for space heating and industrial boilers and in rural areas in Germany and is still being used.

Thus, availability of cheaper primary fuel to industry and rural areas would give fillip to local industry and improve living conditions in rural areas. Processed (Briquettes or

Dried Pulverized) lignite can replace imported coal for cement plants, used in Brick Kilns, and can be burnt in industrial boilers of textile and other industries. Only minor conversion is required to convert from gas and oil. Rice mill boilers and other rural industries can also be converted to this fuel. It should save more than 50% of fuel cost of industries. In cold rural areas like FATA, AJK and Gilgit and Baltistan, Chitral and elsewhere, institutional buildings can utilize boiler based space heating using this fuel.

Electrical Power Demand and Supply

How much Electricity do we need has become a very important yet complicated question in the wake of heavy investment in the energy sector that is being done. There are sceptics and pessimists who are warning that the debt crisis would emerge due to heavy CPEC investments and that economy would ultimately suffer under balance-of-payment issues. Within the government departments, there is now a question circulating around as to how much of energy capacity is enough?

Energy has many elements or components; electricity, Gas, Petrol, Diesel, Kerosene etc. Petroleum products deserve lesser attention because these can be easily imported as and when required and long-term capacity development issues such as storage are not that critical. Gas and Electricity, generally, cannot be imported that easily. Transport infrastructure such as pipelines, specialised terminals and transmission lines are required and take a long time to install. In Chapter 30: How much Electricity do we need, we have developed demand projections taking into account the studies that have been undertaken in this respect. We have also developed a list of projects that are at various stages of development. In the end we have speculated on the demand supply gap or surplus of it as the case may tend to develop.

Load Shedding and Transparency

Intense heat and heavy load shedding has compelled me to write something about it as to why load shedding occurs and how it could have been reduced if not eliminated altogether. The issue is quite controversial and all kinds of advice and accusations are available of which we would like to take a stock of. In all the advice and proposals that are normally given, transparency of the type that is required is generally not included. We will take up the issue of load shedding first and discuss transparency in Chapter 31: Load Shedding and Transparency.

It is generally claimed that GOP did not utilize the available capacity so much so that even IPPs were under-utilized due to the policy needs of reducing circular debt. The data in fact reveals that IPP have been under-utilized. The million-dollar question that needs to be answered is that what amount of load shedding is due to mismanagement and inefficiency i.e. what amount of load shedding could have been avoided with better efficiency and without alleged policy of controlling circular debt and subsidies and thus producing less.

DISCO Performance

NEPRA has released Performance Report for the year 2014-15 recently which gives us an opportunity to have some discussion on the issue. Following broad conclusions can be made as a result of examining the report; none of the DISCO is up to the mark in terms of performance criteria (which we will discuss later in this space); No Company has improved its overall performance (except minor improvements) over the last five years; in fact, there are cases of worsening performance. We have examined in Chapter 32: DISCO Performance if there was some scope for improvement? Of course there is. Efficiency, Quality, Service Level and Profitability should be the yardstick of measuring company performance. In terms of loss reduction, IESCO, FESCO and GEPCO have reached levels, where further reduction may not come so easily as these companies are close to the target of around 8-10 %, although all other companies have a long way to go. With T&D loss reduction, company profitability would improve, as at this moment most companies are paying for a portion of T&D losses from their own pocket i.e. consumer tariff does not cover these losses. With increase in profitability, cash flow would improve and companies would be able to invest out of their own funds for expansion and improvement. NEPRA Report does not deal with company profitability and I have not seen relevant reports, so I would leave it at that.

Reorganization of Electrical Sector

Despite all kind of irresponsible discussion by the TV anchors and apparent openness, there is still lack of transparency and dissemination of correct data and information, based on which objective analysis can be made. Hence, we cannot be the arbiter of this controversy, as to who is responsible. It appears highly unlikely that economy managers would be naïve enough to ignore energy, while on another plane they are allocating a lot of debt resources to the same sector. Blame has to be shared, probably, by both. Or it is the undue centralisation and accumulation of power in a few hands that is responsible. In this article, we will focus on the disintegration of the power sector, which in our opinion has played a major role in the less than optimal performance of the sector.

Power sector has experienced disintegration under the utopia of the IFIs (international Financial Institutions). Their idea was essentially based on reducing the role of government and bureaucracy by bringing in private sector. However, privatization in Pakistan has been limited. If the recent statement of the new secretary Water and Power is correctly reported by the press, privatisation plan has been deferred or shelved indefinitely.

Before restructuring of the power sector, there used to be an intermediary level of professional management, intervening and communicating between government and the operating companies. Now bureaucracy in Islamabad is directly managing the companies and handling the day-to-day affairs. Ministry should instead only be concerned with policy and ensuring that rules and procedures are followed. In Chapter 33: Reorganisation of Power Sector, we examine and posit options on these issues.

Federalism in Electricity

Electricity is largely a federal subject in our constitution with some role for the provinces in small-scale power generation and the sector being under the oversight of Council of Common Interest (CCI), although some confusion persists as evidenced by the formal and informal role of provinces in policy making and investments. Sindh Chief Minister, Murad Ali Shah has threatened or warned HEPCO, SEPCO and WAPDA to pack up or stop the disconnection operation in Sindh. He added that Sindh Government can handle its electricity sector itself. It is not the first time that a chief minister of a province has expressed the desire for autonomy in the electricity and energy sector as a whole. Similar statements have been issued earlier by KPK on issues related to tariff, hydro royalty, autonomy etc. Although self-control and management is a longer-term concern, more practical issues are at stake momentarily. Small provinces are a beneficiary of pooled pricing as we shall see later, but the provincial governments are behaving as free riders by not supporting the elimination and control of electricity theft drives launched by DISCOs. In Chapter 34: Federalism in Electricity we examine the allied issues.

Reviving PEPCO

Over the years, power sector has been restructured and dismembered. It has gone from one extreme of a monolith WAPDA to another extreme of smaller splitted organisations which are not able to accumulate expertise and develop and absorb technologies and skills. Resultantly, there is hardly any major initiative or proposal which does not emanate from foreign consultants and experts tendered by IFIs (International Financial Institutions).We need an accumulator organisation. This can be done by reviving PEPCO which should control DISCOS and GENCOs as well. Although some GENCOs (burning Furnace Oil) will be closed down, many new DISCOs would emerge. It would become well-nigh impossible to control and manage all these organisations from the ministry in Islamabad. Large companies controlling many organisations have similar central organisations even in the U.S., Europe and Japan. In India and Korea, there are integrating organisations of this nature. By reviving PEPCO in a slightly altered form as proposed earlier, the current extreme would be balanced. In Chapter 35: Reviving PEPCO, we examine the perspective and rationale of the proposal in detail.

Regulatory Reforms

Two actions have been taken by the Ministry of Water and Power, of late causing quite some stir among the stakeholders. First, NEPRA has been put under direct administrative relationship if not control of MoWP, in place of the Cabinet division, which may not be as consequential as is being perceived by the public. Second, a summary has been moved to CCI (Council of Common Interest) to reduce the powers of NEPRA and bring about some associated changes. While the first step of administrative reporting appears to be catering more to psychological needs of the power ministry bosses, the second step is of more serious concerns and consequences.

We have taken a stock of the proposed changes in Chapter 36: Regulatory Reforms in a non-partisan manner and undertake an objective and cool analysis. Some of the proposals are meant to remove barriers to entry and initiation of the market process, and bringing an element of planning into the system, which perhaps everyone would support. The other proposals are more or less aimed at reducing the role and power of NEPRA, which may be contentious requiring more consultation and expert input for bringing some balance.

Alternative Approaches to Smart Metering

Smart meters are new generation of electric and gas meters having a memory chip and an electronic communication device integrated with a communication network like GSM, GPRS, and Broadband , Wi-Fi etc. They can do many things: send hourly consumption data to utilities, monitor demand and connected devices, change allowed loads and disconnect and reconnect from the computers centrally installed in utilities. They eliminate need of errand meter readers and enables disconnection of defaulting consumers, and reconnecting once default is removed, without the need to send a lineman who is often found to be corrupt aiding and abetting in electricity theft. Just as Income Tax reforms have eliminated direct contact of taxmen with the taxpayers considerably reducing corruption and nuisance, smart meters may prevent corruption by eliminating the need of utility employees running around the lanes and doing all sort of bungling. DISCOs would be able to monitor theft at higher professional level and take appropriate measures. In Chapter 37: Alternative Approaches to Smart Metering, we have undertaken a critical evaluation of the proposed scheme and have made some submissions with respect to some better and more realistic approaches.

CASA- a Liability

CASA is a project planned for importing electricity from Tajikistan and Kyrgyzstan through a 750 kV HVDC transmission line 1000 km long passing through some of the most difficult terrain. It will bring 1000-1,300 MW electricity into the system. CASA was conceived more than ten years ago at a time when oil prices were high, foreign investment was scarce and electricity deficit was on the horizon. It was kind of viable at oil prices of the order of 120 USD per barrel and more. It was a short-term solution to be implemented within two years.

Independent economists have been cautioning against mounting debt and have issued forecasts of balance-of-payment difficulties and of other associated issues. Recognizing that investments have to be made to solve the problems such as energy deficit, the least that can be done for tackling these issues is to invest in projects, which have good pay-offs.

We have examined the plausibility of the project, in Chapter 38: CASA-a Liability? in the evolving power surplus position, security and safety risk due to continuing instability in Afghanistan, emerging electricity economics in the context of low oil prices at-least in the mid-term scenario, and possible negative foreign exchange impact among others and suggest measures to improve the project, and postpone the project if not cancel it altogether. In fact, there is a need of conducting a review of a number of other projects, which have lost relevance in the changed circumstance.

The New K-Electric Tariff

NEPRA has recently announced new Tariff for K-Electric, which will be valid for the next seven years. The last tariff determination was made in 2009, which in fact was almost a renewal of the tariff originally issued in 2005 immediately after its privatization. In this article, we would attempt to appraise the tariff determination in terms of its strength and weaknesses and make some recommendations for possible improvements. Overall,

NEPRA determination appears to be all right, provided there are no issues in data and calculations. K-Electric does not appear to be sure of it and asked NEPRA for working details. K-Electric also thinks, that required investment may not be attracted under the instant determination. Consumer groups are not happy either, but it is not clear why? Both parties are preparing to request for a review. The axiom goes that when both parties cry, the arbitration or decision is correct and balanced. In Chapter 39: The New K-Electric Tariff we examine the allied issues.

High Relending Charges

Interest cost and rates play an important role in determining the final prices in energy sector, which is highly capital intensive. In Pakistan, public sector still plays an important role in the sector despite the advent of IPPs. GoP serves as an intermediary in all the borrowings from the international bodies made by public sector projects, possibly due to the sovereign guarantees issue, but for this it charges a heavy margin: borrowing at 1 or 2 % and relending at about 14-16%. This may be a way of cross subsidizing the social sector projects, but creates a lot of difficulties for energy sector projects. In Chapter 40: High Relending Charges, we make some submissions for correcting this situation.

Nuclear Power

In 1970s, when nuclear bomb was being made, there used to be statements about some 25-30 nuclear power plants of 25,000 MW capacity were to be built by 2000 to meet the projected demand. Nuclear Power was probably the cheapest of all. There was great hope and expectation from nuclear. This has not happened. There were several nuclear accidents; Three Mile Island in the U.S., Chernobyl in Russia and Fukushima in Japan. Most Europe has vowed to close all the existing nuclear power plants at the conclusion of their normal life. Germany closed all of its nuclear plants but later reversed the decision in favour of normal closure. Safety risks, terrorism, waste management, bomb proliferation and bad economics have pushed nuclear power at the bottom list of priorities of nations, if not a total closure of this option. Westing-House, probably, the largest nuclear reactor company has been bankrupted and more may follow. No nuclear reactor has been built in the U.S. after 1980.

Up to now, a general impression has been that the nuclear power is a competitive energy source, if not the cheapest, which has remained a contentious issue in the U.S. Debate has been around for hidden subsidies and internalizing or lack of including externalities such as accident costs etc. Supporters of nuclear power manage to prove that it is competitive, if not the cheapest and opponents also manage to prove their case that it is expensive and uncompetitive. An additional factor that has come up recently is very high capital cost of the safer Gen-II nuclear power plants costing in excess of 5000 USD per KW. Pakistan is acquiring ACP-1000 nuclear power plants which are a variant of AP-1000 Gen-III reactors – Westinghouse type reactors. In Chapter 41: Nuclear Power, we will do a survey of similar projects in other countries with a view to examine the cost structure there and compare it with potential costs in Pakistan.

In Pakistan, after the initial plan of 25,000 MW in 1970s, another Energy Security Plan was made in 2005, which included a 8000 MW nuclear power capacity by 2035.Three power plants of 330 MW each have already been commissioned and another one is under implementation. Karachi KANNUP has completed its life, although, its life is being extended. This is being considered by many as to be too much of a risk for a small capacity of 100 MW. Heroism should have its limits.

Two Nuclear Power Plants of 1000 MW have been approved. Some Construction on one of the two plantsKPK-1had already been initiated. Reportedly, there is no progress on KPK-1, due to financial problems. GoP could not mobilize its share of Equity. GoP has other priorities. It would finance Bhasha rather than a nuclear power plant.

There are acute siting issues.KP-1 attracted a lot of controversy. And there are siting issues for other power plants. Balochistan coast offers an opportunity but has been marred by Tsunami considerations and political and Law and order situation there. According to many, it would be too dangerous to site nuclear power plants on the only river that is available to us. One would not like to take any risk in that respect. No provincial government is likely to agree with such a proposition.

Nandipur Project

Nandipur power plant project, in 2015, attracted a lot of media attention. An analysis of the arguments that were put forth indicates that quite a few people were either confused or were ill informed. In Chapter 42: Demystifying Nandipur Project Controversy, we have put together the relevant facts and figures and do some analyses so that a reasonably clear picture emerges and doubts and fears are removed from public specter. I have to clarify here that I have not been involved with Nandipur project and I can thus claim to present an unbiased and neutral account. Although some of the data may be drawn from government public documents, most of it is, however, from NEPRA, which can safely and hopefully be considered a source of unbiased and reliable information.

The impending questions that this chapter attempts to answer are: has CAPEX reached an impossibly high and unaffordable level? How does the CAPEX compare with other comparable plants? Was it more expedient to shun the project rather than revive it? Would there have been much greater loss in shunning the project as opposed to the additional cost that has been paid for in revival, as we would demonstrate later in these passages? What would be the cost of generation as compared to other comparable plants? And finally what are the prospects of running the plant without undue interruption?

2. Strategic Directions in Energy Sector

In this chapter, we intend to paint a broad picture of the next 20-25 years of the energy sector including oil, gas, coal and electricity. We will try to speculate on as to how demand, supply and prices would play out so as to draw some policy conclusions and recommendations. In order to be able to do that, it is necessary that we consider how the energy sector has behaved at least in the recent past.

Energy consumption in Pakistan, in the last 5 years, has grown at a rate of only 1.6 % due to low economic growth and supply constraints. In the same period, electricity consumption grew only at a rate of 2.3 % due to the obvious reasons of supply deficit, as against a rate of 4-5% long-term historical growth rate and a rate of 7% p.a. required for optimum economic growth of the country.

Sector wise energy consumption as a percentage of total consumption stands at, 24.5 % for domestic, 35.4% for Industrial and 32.4% for transport respectively. The averages hide good and bad and high and low; for example, domestic sector energy consumption grew at 4.2% and transport at 3.1%. Within transport sector, petrol consumption has grown at a phenomenal rate of 19.8 % as against stagnant demand for diesel. We used to be self-sufficient in petrol sometimes back, but now we import petrol along with diesel. Rising petrol consumption is largely due to high imports and production of vehicles, more so of motor cycles, although, partly lack of CNG is responsible for higher petrol consumption. High proportion of petrol in overall transportation fuel mix has some policy implications in terms of pricing that would be discussed later in this space.

Overall, oil consumption continued to grow at a rate of 5% p.a., occupying a share of 33% in overall energy mix. Gas, despite stagnation in supplies, continues to occupy the highest share of 37.5% in the overall energy mix although its share has come down from 43.9 % in 2009, possibly replaced by oil, share of which increased to 33% from 27.9%. Lower oil prices and gas supply constraints seem to be responsible for the change in the energy mix. Oil imports have increased from 18 million tonnes to 21.7 million tonnes. Although, in value terms, due to oil price decrease, oil imports totaled at 13 million USD in 2014-15 as opposed to 10 million USD in 2009-10. Trends of the recent past can be summarized as follows:

1. Oil exploration and development (E&P) sector has done quite well. Crude Oil production has increased from 65,866 Barrels per day in 2011 to 97,000 barrels per day in 2016. This is a growth rate of 7.8 % p.a., as opposed to almost stagnation in the ten prior years of 2001-2010 at 63-65,000 barrels per day. This has come on the back of almost a breakthrough in KPK and increased efforts in Balochistan. KPK crude output has tripled from a mere 5 million barrels per annum to 16 million barrels, a growth rate of 25% over the last 5 years. KPK's output in gas sector has also shown similar trends as we shall see later. Most energy components have either stagnated or have grown at about 4 % per year. Thus crude sector has grown at twice the average rate .It would be a good omen, if this rate is maintained in future as well. Chances are that it will, as the sector is being reorganized under 18th amendment and provinces are taking more interest not only in control but also in terms of actual effort.

2. Gas production, on the other hand, has stagnated at 4 bcfd while the demand stands at 8 bcfd, as per the data collected by Ministry of Petroleum and Natural Resources (MPNR) . Consequently, all end-users have suffered gas load-shedding including domestic, commercial, industrial and power sectors. Gas output has increased in KPK only at the rate of 11.8 % p.a., so much so that not only KPK is self-sufficient in gas, it is in surplus, production of 131 bcfd as opposed to a consumption of 74 bcfd. KPK now demands that gas fired power plants be installed there on the new output. Gas reserves

Table 1 Energy Components Growth Rates (%p.a.)

Growth Rates(%p.a.)	2003-2008	2010-2015
Oil	1.2	5
Gas	11.8	-1.5
Electricity	6.8	2.9
Coal	26.2	1.6
Motor Spirit	5.9	7.5
HSD	3.3	3.9
Furnace Oil	0.6	0.4
Market Share (%)		
Motor Spirit	11.7	21.35
HSD	29.94	33.44
Furnace Oil	33.72	41.53

Source: Energy Yearbook -2015

in Sindh and Balochistan have stagnated. Prospects of further discoveries and investments have come down with the reduction in international Oil and Gas prices. Shale and tight gas reserves have been proved, however, production cost has been estimated at 12 USD per barrel - more than twice the existing local gas production cost and also of RLNG. Besides, water requirements are excessive in Shale output and also the prospects of water pollution indicate that time of Shale gas has not arrived yet for Pakistan. Thus in the near future, dramatic or major increase in gas output may not be expected. It would not be a bad idea to rely on imported LNG for the next ten years in the circumstances that have been indicated.

Introduction of RLNG has made some difference. One RLNG terminal (Engro) is already operating while another is expected to come on line this year. 3 more terminals are at various stages of planning. One RLNG terminal adds a capacity of 4 bcfd, thus 12 bcfd should be added in the next 3-5 years. It may be of interest to note that a 1,000 MW power plant consumes about 100 mmcfd of gas. Transmission may create bottlenecks in mid-term future, unless a main transmission trunk line from South to North is implemented for which negotiations have taken place with Russians. There is a local pipe-line installing capacity and capability as well which is cheaper. Even, there are underutilized pipe mills as well.

3. In the **power sector**, electricity consumption grew at a high rate of 6.9 % p.a., in the period (2003-2008) when GDP growth was high at 6.1% p.a. The growth rate of electricity consumption came down to 2.9% p.a. in the period (2010-2015) when the GDP growth rate had been lower at an average of 2.5%. Installed capacity in the year 2015-16 was 23,179 MW while gross generation stood at 100,659 GWh. Hydroelectricity provided 34%, Gas 27 % and Oil 33% of total generation. Renewable Energy share was less than 2%.

The installed capacity has increased by only 6,000 MW over the last 17 years in this century. Output has stagnated at 18,000 MW in summer and 12,000 MW in winters as against an installed capacity of 23,179 MW by 2015-16. A shortage of 6,000-8,000 MW has been recorded in the system. However, by 2018, 10,000 MW would be added which is expected to eliminate the shortage. In fact, some surplus has been indicated for the initial years. Thus by 2018, the installed capacity is expected to increase to 28,000 MW.

Table 2 Pakistan's Power Generation Mix 2015-16

Source	Installed Capacity (MW)	% of Installed Capacity	Generation (GWh)	% of Generation
Hydel	7,116	30.70%	34,272	34%
Furnace Oil/High Speed Diesel	4,321	18.64%	33,296	33%
Gas	10,097	43.56%	27,289	27%
Coal	150	0.65%	148	0.15%
Nuclear	650	2.80%	3,854	3.83%
Wind	306	1.32%	786	0.78%
Solar	400	1.73%	207	0.21%
Others	139	0.60%	807	0.80%
Total	23,179	100%	100,659	100%

Source: Power System Statistics 2015-16

Thus if economy grows at the high rate, 20,000-25,000 MW of additional capacity would have to be added i.e. 2,000 MW per year. It means that there is to be no respite and no resting in between. Same tempo would have to be maintained, as has been the case in the last 4 years. How should this additional capacity be proportioned is the question energy planners have to deal with.

Almost 50% of the additional capacity in the period 2018-2028 is to come out of Hydro: 6,000 MW Bhasha, 4,000 MW Dasu and 2,000 MW of private sector projects. Thar coal should at-least get another 5000 MW. Despite large resources of 185 billion ton in short to medium run, an upper limit of 8,000-10,000 MW has been indicated due to other constraints such as that of water. RLNG capacity of 5000 MW may be added to be located in Punjab. May be another 2000 MW in Sindh and KPK on local gas. Wind and Solar of 2,000-3,000 MW may also be accommodated. This arithmetic appears to be realistic. For Dasu, World Bank financing is available, and present government intends to start Bhasha either under CPEC or under self-finance by WAPDA. Inputs (fuel side) for both Thar and RLNG are in place or in the pipeline. Thar may compete with imported coal power plant proposals. Conversion of oil-fired power plants would be a tough competitor for Thar. Government of Sindh has to become more cooperative and act as marketers than sitting on the resource. It is a competitive world. They should attract Government of Punjab to put up their equity in financing and owning Thar based coal power plants, as has been suggested by Khwaja Asif, Federal Minister of Water and Power. According to him, and rightly so, it would promote national cohesion. While hydro plants may take 7-10 years to complete, electrical output in the intermediate period (2020-23) can come from the proposed RLNG power plants. Another 2,000 MW may also come from nuclear. Problem with nuclear is of high initial capex - in the cost of one nuclear power plant, 4 gas fired power plants are installed.

Figure 1 Future Demand and Supply

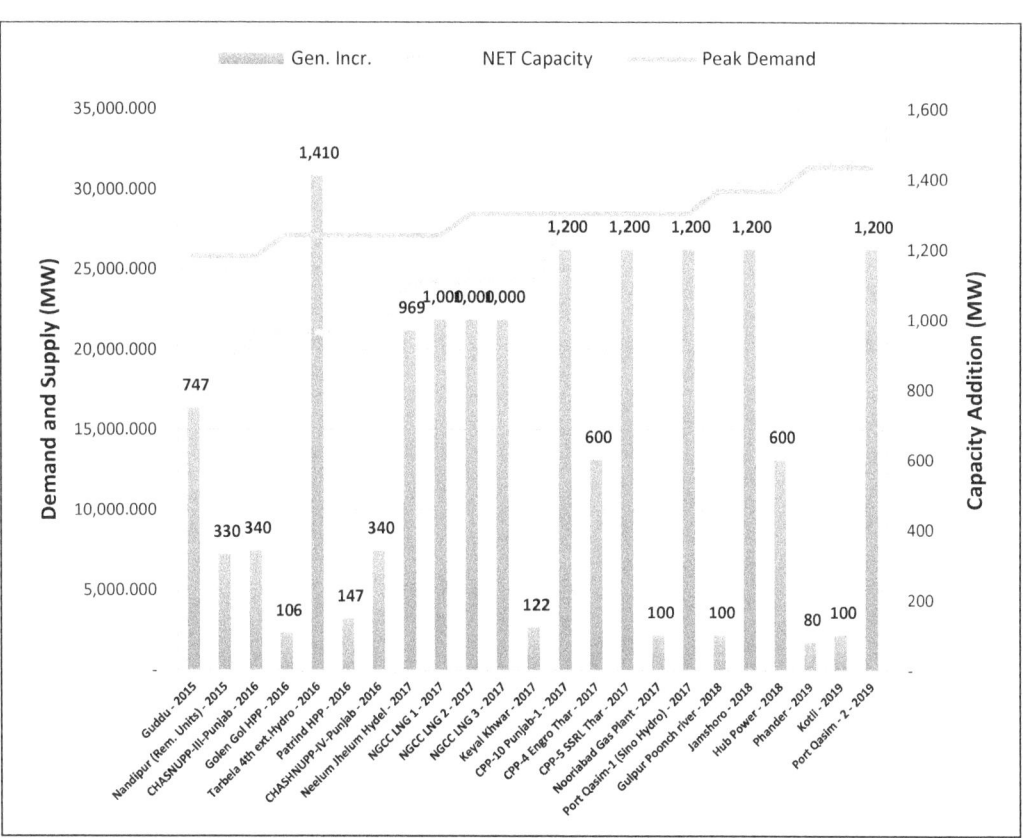

Source: NTDC, Planning Commission & MWP

Figure 2 Future Demand and Supply 2014-35

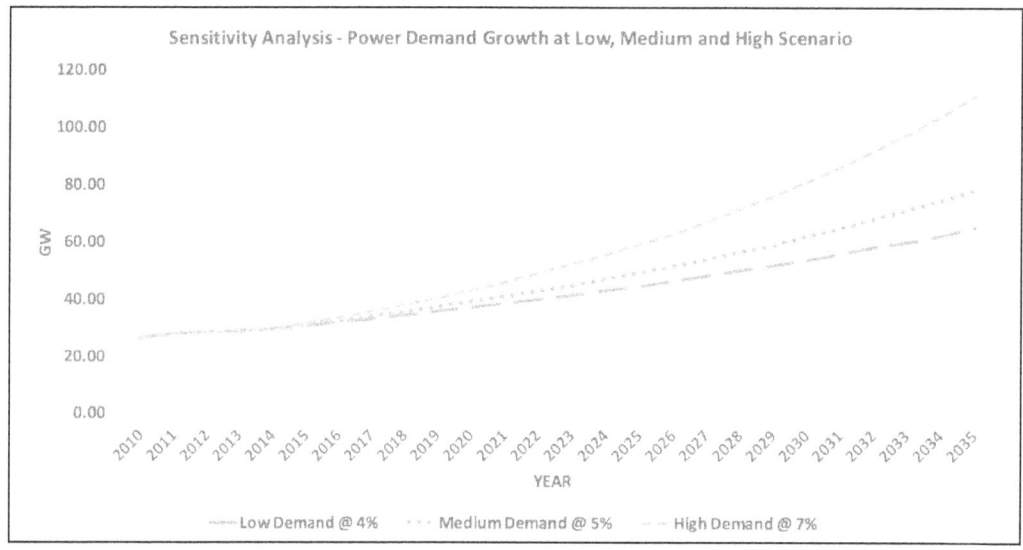

Source: JICA Report

Figure 3 provides the NTDC power development plan as developed by NTDC. As seen, coal, LNG and hydro plants constitute major proportion of the incoming power plant additions.

Figure 3 Power Development Plan up to 2035: Demand and Supply Balance (MW)

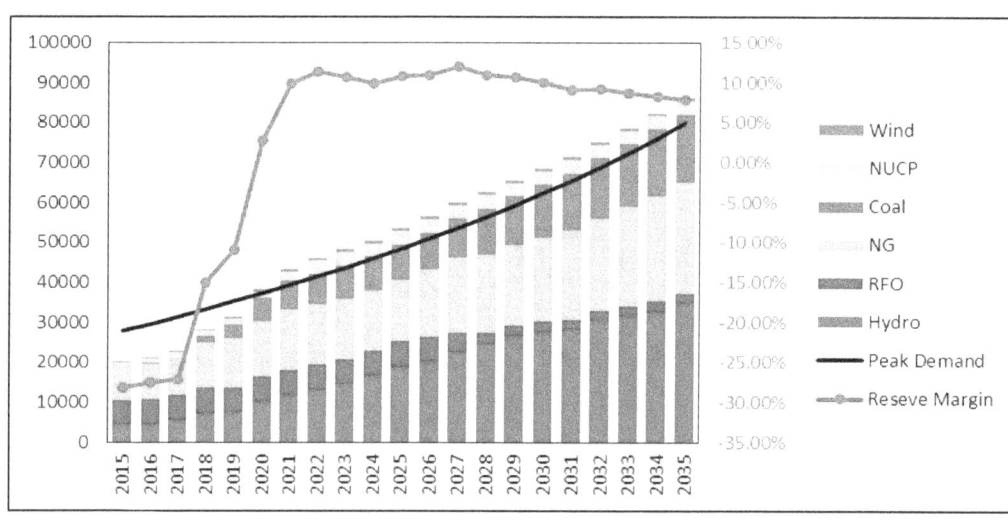

Source: JICA Report

The period 2028-38 appears to be problematic, if current technology and approaches of central generation are kept in mind; addition of 45,000-50,000 MW in ten years, 5000 MW per year, would be required. However, the question is if we would we be that efficient in that period. Neither hydro nor coal alone can be enough, due to other issues if not due to lack of physical resource size. Perhaps, that would be the time for distributed generation based on solar and to some extent wind. In that period, storage technology would be in place as well. Even nuclear of 8,000 MW may be feasible by then. Kalabagh dam may be revived, as with time sanity may prevail and water scarcity problem may be more visible to the provinces by then. Local Shale Gas may also emerge by then. And oil prices may return to 100 USD per barrel plus.

Keeping these future projections in mind, indigenous capacities and capabilities may have to be developed which at present are either not there or are being underutilized. The current period has been of crisis requiring faster speeds of implementation, which meant almost sole realization of imported inputs. In future, emphasis may have to be on development of local capacities and capabilities. A power plant design and manufacturing plant should be installed under Chinese assistance under CPEC. Planning capacities would also have to be developed. Foreign consultants and international agencies, which fund them have their own ideologies and priorities. Japanese don't like nuclear and thus excluded nuclear from their considerations and recommendations, others do not like coal, and still others bring in political issues in financing hydro. An integrated approach to energy planning is required with the objective of cost-minimization. Earlier attempts by Planning Commissions did not succeed in establishing a credible system in this respect. Let us hope that existing and future schemes succeed. Furthermore, the reforms and restructuring of the energy sector must continue bringing in privatization and market forces into the sector.

Although, it is difficult to speculate on future, especially, the energy prices, consensus seems to be emerging that oil and gas prices would regain their high level in a matter of next ten years i.e. by 2025.The honeymoon of RLNG would last only that long though high efficiency and flexibility of operation would keep gas fired power plant relevant n the next twenty years. But too much of power plant capacity on RLNG may not be advised. Coal is attracting a lot of opposition internationally due to Climate change issues. Some kind of international restrictions resulting in higher prices of traded coal may emerge. Long-term future of imported coal appears to be dim. Local resource argument may work in favour of Thar coal. Hence, there does not appear to be much of a justification installing any more power plants on imported coal. Studies should be commissioned on the technical and economic feasibility of converting imported coal plants on Thar Coal in the medium term. Foreign exchange constraints may also require us to do so.

Petrol consumption has been increasing at a phenomenal rate. Indeed, transport sector is totally dependent on Oil. Transport sector's share in overall energy consumption is 30%%. Road pollution is becoming unbearable, even in Islamabad. Not much attention is being paid to this aspect. Many countries have started developing plans for deploying EV (Electrical Vehicles). Pilot Projects may be initiated in this respect for public transportation (Buses) in polluted and congested areas.

3. What is the Right Energy Source?

There is no right energy source: all energy resources have one limitation or the other. Solar is only available in the day, while hydro and wind are available only in summer. Coal has environmental issue and there are other limitations in Thar limiting production possibilities to 10,000 MW, Furnace Oil is expensive and Gas resources are depleting and imported LNG is expensive and will cause foreign exchange drain. Nuclear has a large number of problems and limitations.

All cannot be rejected on negative grounds, and all cannot be inducted 100 %. There is to be an optimal mix. We have provided two comparative tables in this chapter: One compares qualitative characteristics and the other provides some quantitative prameters. After perusing these tables, following policy conclusions can be made:

1. Solar PV has come of age and is now the cheapest source available from a variety of sources. IFI's and other countries are available to finance renewable energy projects. It is available almost everywhere in the country. For grid connection, especially, for a week grid that we have, there are some limitations. It is said that up to 10% of Renewables (Solar and Wind) can be integrated under current conditions. This gives a figure of about 2,000 MW in the short term and another 3,000 MW in the next decade, taking the total to 5,000 MW. In the long run as more technical progress is made and storage becomes available and Solar PV cost comes down further, Solar PV has a potential to go up to 20 % or even 50%. Preference should be given to distributed generation: small 20-50 MW plants distributed and dispersed throughout the country.

2. Wind resource is located mostly in Sindh. Some potential has been indicated in Northern areas but could not be confirmed yet. Punjab may have some limited resource in Kallar-Kahar region and Rajanpur. In Western Balochistan, there is a large and scattered resource. In Chaghi and around, there is a large Wind resource, especially, around the junction area of Iran, Pakistan and Afghan border offering opportunities of cooperation among the three countries.

Currently, only Sindh resource appears to be exploitable. In Balochistan, small Wind-Solar projects can be launched for individual or combinations of population cluster. Wind Power has many characteristics common with Solar. Solar-Wind combination can counteract each other's problems to provide almost a year-round availability. 30% of Renewable quota can be offered to Wind Power and 70% to Solar. In Wind Power, break through is not expected. It is a mature technology and one should expect only gradual improvements.

3. Hydro Power has been on agenda since 1960s. A potential of 50,000 MW is indicated. There is a project portfolio of 30,000 MW on which work has been done and projects are at various stages of planning. Much time was wasted earlier in debate over Kalabagh dam, and later IFIs wasted and procrastinated on Bhasha. Hydro used to be the cheapest energy source, until recently, its tariff was under one Rupee. This was due to older and depreciated investments. All new projects are going to have a COGE (Cost of generation) of 8 Rs or more per unit, higher than Solar, Wind and even coal.

Hydro has the highest capacity factor among the renewables of 40-45%. It gives electricity mostly in summers. One has to install, additional matching capacity for winters. Only nuclear industry people have in a way opposed the hydro presenting the argument that hydro is as capital intensive as nuclear if capacity factor is taken into account: hydro has a CAPEX of 2.5 Million USD/MW with a capacity factor of 40-45%, while nuclear has a CAPEX of 5 Million USD per MW with a capacity factor of 80-90%. They have a point, but there are other problems with nuclear that we would discuss shortly.

Another popular misunderstanding about hydro is that it brings water as well which is not correct. Only Tarbela and Mangla have water storage, others are simply power plants and small reservoirs that are made to balance the water flow variations. Bhasha, KalaBagh Dam and Munda will be the storage dams as well besides producing electricity. Main problems in hydro are land acquisition, difficult logistics, and difficulties in construction and long gestation of time. That is why, hydro power could not be implemented despite general support and early start. Remoteness is another issue, not only causing problems in construction logistics, but problems in transmission and heavy investment thereof.

The advantage being that large power plants (which are although being criticized by many) are installed which are highly versatile and can run in almost all modes, base load, intermediate and even peak, and requires only seconds to ramp up the generation provided the extra capacity is available. Normally, this is not the case.

Hydro Power has remained on national agenda since 1960 and a main milestone of Tarbela Dam (3,478 MW) commissioned in 1976. Present capacity has come up in stages, as the project stated with a smaller capacity. Its capacity is being further expanded by 2000 MW through two World Bank funded projects, Tarbela IV and V. Total Installed capacity of Hydro is 6902 MW. Projects include Ghazi Barotha (1450 MW), Mangla (1000 MW), and other smaller plants. A new but small power plant Patrind in AJK with 147 MW capacity has been commissioned these days (16th June, 2017). Neelum-Jehlum 968 MW in AJK is about to be completed.

There is some progress on Bhasha Dam (4500 MW), which has been inaugurated many times over the last ten years. Fortunately, most of the preliminary operations including land acquisition has been completed and more than Rs .100 billion have been spent already. IFIs are reluctant to finance Bhasha due to what they called disputed territory. WAPDA had prepared a proposal to initiate and finance Bhasha through local resource by separating dam and power portion, the latter to be built under IPP mode. Admittedly, it would not be easy to finance such a capital intensive project through local resources, but one may have to willy-nilly undertake it, if nobody else is prepared to cooperate. Chinese are reportedly interested in a very ambitious proposal of taking over the whole Indus involving, Bhasha, Bunji, and others. This would add to a total of 20,000 MW involving an investment of about 50 billion USD, almost another CPEC. If one adds to this, the parallel negotiations pushed by PAEC that are sheepishly going on the subject of several nuclear power plants, another 50 billion USD is added.

Chinese want to take over some existing projects as well and want to operate and optimize the whole Indus cascade. This is a tall order and involves many risks and potentialities. It is highly unlikely that a decision in this respect can be taken under the current political transition. If the nuclear and Hydro are combined, nothing seems to be feasible. It is hoped that something comes out for Bhasha dam at the earliest opportunity on which there is a national consensus unlike KalaBagh dam. We need Bhasha more for water than for electricity, for the latter there are many other options. For storage dam, there are Bhasha and KalaBagh only and some other smaller possibilities.

Another lesser known project of importance being quietly built is DASU of 4320 MW; the project would be built in two stages. First stage is of 2160 MW construction of which has recently started with EPC contract going to a Chinese company. The project is expected to be commissioned by 2021.WAPDA is implementing the first stage, it is expected that the second stage, in which only Power Plants would be built may be implemented under IPP mode. It is also a storage dam and thus has importance from water security point of view.

Table 3 Electric Power Installed Capacity vs. Water Storage

	MW	Storage(Km3)
Tarbela	3,478	13.69
Mangla	1,000	9.12
DASU	4,320	1.41
Bhasha	4,500	7.9
KalaBagh	3,600	7.52
Munda	740	1.59
Total	17,638	41.23

Source: WAPDA

4. Oil based Power Plants: Perhaps most of the power supply woes in Pakistan has been due to an undue share (36.8 %) of Furnace Oil based power plants with an installed capacity of 4,304 MW .They overlap with Gas power plant capacity(10097 mw), as most of these are dual-fuel plants. Gas not being there, Furnace Oil is used to run the dual fuel gas plants. Furnace Oil being expensive, it is not usually supplied by GoP to power plants. Neither, there is foreign exchange, nor money, as subsidy requirements increase with more generation on Furnace Oil. Hub Power was installed and commissioned with a lot of fanfare in 1995 as a first IPP. It remained under- utilized for a long time because, it came last in the merit dispatch order. Later on as the electricity deficit increased, all power plants had to be run. Being the most efficient one, thermally and operationally, it was later utilized to almost full capacity. It is being closed down on completing its normal life .Hub Power is in its place installing a Coal power plant based on imported coal of 1300 MW, which is not being liked by many, which will add to foreign exchange drain and steal market share from Thar.

Oil fired power plants are easy to install. Oil can be imported easily as well. There is no other logistics required such as elaborate terminals in case of LNG or even coal. Smaller power plants, e.g., I.C. Engines up to a capacity of 150-200 MW can be installed by putting skid-mounted sets of 10-15 MW each. Their CAPEX is also low, under one million USD per MW, although, in Pakistan, it has been higher due to the usual issue of capital-padding and corruption. These power plants ramp up time is very small, so they are used only for limited time for peaking purposes in most countries. Due to little utilization of less than 500 hours, their capacity charge is high and COGE can go to as much as 30 cents in many countries. In our case, due to shortages, such plants have been operated as base load or intermediate load plants. When there is acute shortage, there is no distinction.

Large steam turbine type Oil fired power plants have been built mostly when Oil was cheap or such power plants have been built by oil producing countries like Iran, Saudi-Arabia, Kuwait and the U.A.E. Pakistan would probably be among the very few countries that might have installed oil based capacity despite not having oil. And kept adding. To be fair, these were dual-fuel power plants, which are being run on oil because gas or cheap gas is not available.

5. Natural Gas: Natural Gas has come to lime light for a variety of reasons. Among fossil fuels it is the lowest polluting and has the lowest GHG foot-print. Shale gas has been discovered in the U.S. and large gas finds have come up such as in Mozambique. High investments have been made in LNG infrastructure and along with Oil prices, LNG prices have come down as well. Classically, there used to be a ratio of 80% between the price of oil and LNG.LNG has come down to a lesser ratio. However, this is temporary. LNG/Gas prices are expected to go up again as oil goes up to USD 100 per barrel in less than a decade.

However, the most sustaining competitiveness of Gas has emerged from increasing efficiency of NGCCs (Natural Gas Combined Cycle Plants).It has achieved an efficiency level of 60%,which is the highest; higher than 30-35 % of steam plants ,although coal power plants are also coming to 45% level or slightly more.

In Pakistan, fuel costs of less efficient power plants running on gas is around Rs.5.00, because locally produced gas is cheap. The fuel component of RLNG plants would also be around the same at Rs.5.00 although RLNG is twice as expensive as the local gas. The reason being the unprecedented efficiency of NGCC. Three power plants of RLNG have been acquired, two by Federal Government and one by Government of Punjab. An additional RLNG power plant has been approved recently, also for Punjab. There are proposals for adding RLNG power plants in Sindh and KPK (on locally produced gas in KPK).

Locally produced gas is getting depleted by the day with no major replacement of reserves. The forecast is that it would be depleted within a decade, unless LNG is brought in to take up the load. LNG is expensive costing almost twice than the local gas. The right policy would be to conserve local gas for house-hold purposes, and allocate RLNG for all others. More or less this is being done. But, one should not go too far in LNG enthusiasm, for LNG will again be expensive soon enough than it is being thought.

6. Coal: Coal is the most abundant energy resource in the world. 41 % of world electricity is being produced on coal. In the U.S. coal and Gas market shares are equal at around 31-32% each. It is considered to be the dirtiest as well from mine to shipping to power plants. It emits Sox, NOx, particles, dust, and Hg, besides creating solid waste management issues. There is a wide disbelief about clean coal. Because of high coal usage and its high GHG foot print, it is being roundly opposed in the world. FIs have stopped loaning for coal projects. Even China and India have announced curtailing coal.

Coal power has been popular due to many factors; price, availability, high capacity factor and base load characteristics. In Pakistan, imported coal power plants have been installed in Sahiwal and Port Qasim; the former has started functioning as well. Another power plant has been approved for Hub Power to be located in Hub, Balochistan. All are of the same capacity and Type (1300 MW, 39% Efficiency).The plants have been criticized for higher CAPEX (USD 1.45 Million USD vs 1-1.1 typical), and huge coal imports requirement. GoP does not think so and argues that it was the only fast track option available. This has been proved with early commissioning of Sahiwal Power Plant while Engro Coal power plant in Thar is at-least two years away, despite an early start. There were political uncertainties and difficulties as Government of Sindh was involved in Thar which had to provide infrastructure. It could have gone slow, but apparently has not. Delay in Engro is due to mining and Engro's financial problems which delayed financial closure.

It used to be said that Thar coal can sustain a capacity for 50,000 MW for two centuries or more. But recent studies have indicated some limitations. It is said that Thar cannot sustain more than 10,000 MW due to water availability issues. Coal and nuclear both consume a lot of water which is an argument usually made against these two resources among other reasons. However, some 7000 MW of coal projects are at various stages. Some people in Punjab want to install more coal power plants in Punjab which has been opposed by Ministry of Water and Power and the Planning Commission. It is hoped that they would not insist on their demand. Perhaps, they would be satisfied with more RLNG plants in that province for the time-being.

7. Nuclear Power: In 1970s, when nuclear bomb was being made, there used to be statements about some 25-30 nuclear power plants of 25,000 MW capacity were to be built by 2000 to meet the projected demand. Nuclear Power was probably the cheapest of all. There was great hope and expectation from nuclear. This has not happened. There were several nuclear accidents; Three Mile Island in the U.S., Chernobyl in Russia and Fukushima in Japan. Most Europe has vowed to close all the existing nuclear power plants at the conclusion of their normal life. Germany closed all of its nuclear plants but later reversed the decision in favor of normal closure. Safety risks, terrorism, waste management, bomb proliferation and bad economics have pushed nuclear power at the bottom list of priorities of nations, if not a total closure of this option. Westing-House, probably, the largest nuclear reactor company has been bankrupted and more may follow. No nuclear reactor has been built in the U.S. after 1980.

In Pakistan, after the initial plan of 25,000 MW in 1970s, another Energy Security Plan was made in 2005, which included a 8,000 MW nuclear power capacity by 2035.Three power plants of 330 MW each have already been commissioned and another one is under implementation. Karachi KANNUP has completed its life, although, its life is being extended. This is being considered by many as to be too much of a risk for a small capacity of 100 MW. Heroism should have its limits.

Two Nuclear Power Plants of 1000 MW have been approved. Some Construction on one of the two plantsKPK-1had already been initiated. Reportedly, there is no progress on KPK-1, due to financial problems. GoP could not mobilize its share of Equity. GoP has other priorities. It would finance Bhasha rather than a nuclear power plant.

There are acute siting issues. KP-1 attracted a lot of controversy. And there are siting issues for other power plants. Balochistan coast offers an opportunity but has been marred by Tsunami considerations and political and Law and order situation there. According to many, it would be too dangerous to site nuclear power plants on the only river that is available to us. One would not like to take any risk in that respect. No provincial government is likely to agree with such a proposition.

The chances are very bleak for nuclear power in this perspective. High investment (5 Billion USD per Plant) and high energy cost (Rs 10-12 per unit) and other problems make it highly likely. Many people are opposing the CPEC projects due to its repayment issues and foreign exchange implications. Nuclear would multiply it. Nobody has a genuine interest in it except the nuclear industry itself.

Concluding as said earlier, there is no panacea. All energy sources have positives and negatives. A composite optimal mix with the objectives of least investment and cost of generation has to be worked out and implemented with adjustments. Optima does not always work. Supplier and investors preferences also matter in a country which has not been able to attract FDI for a long time. Local resources are to be preferred even if costing a little more to avoid foreign exchange drain, but should not be a license to promoters and investors to indulge in undue profiteering and capital padding. Coal and other fossil resources are on the way out and Solar is the future. The future, though, has not arrived yet. Let us slide into the future gradually and not jump into it to get hurt.

Comparative Characteristics of Various Energy Sources	
POSITIVES	NEGATIVES
SOLAR PV	
Cheapest & Affordable (5 cents)	Available only in the day and varying
Low upfront investment (1-1.2 Million USD/MW)	Space Intensive
No Fuel, energy Security	Grid Connectivity Issues
Available Every Where, Not Localized	Low Capacity factor (17-20%)
Fast Implementation	Additional matching capacity and investment required or storage
No Pollution, No GHG	Dust and pollution affects output
Distributed Generation	Some Water reqd for washing and cleaning
International Competition	Not for Base Load
International Support and Ease of Financing	
WIND POWER	
Cheap and Affordable (6 cents)	Only Seasonal Availability but throughout the day and varying
Low upfront investment (1.2-1.5 Million USD/MW)	Localized, available in Specific regions
No Fuel, Energy Security, Foreign Exchange Savings	Grid Connectivity and Transmission Issues
Fast Implementation	Space Intensive
No Pollution, No GHG	Low Capacity factor (30-35%)
International Support and Ease of Financing	Additional matching capacity and investment required or storage
	Not for Base Load
NATURAL GAS and LNG (NGCC)	
Cleanest among Fossil Fuels	Local Gas Resource Depleting
Possibly Cheapest among Fossil fuels (5-8 cents)	Imports of Expensive LNG

High Availability & Capacity Factor (60-80%)	Price Variability and risks
High and increasing Efficiency of NGCC(60% +)	Supply and Transportation Risks
Versatile, All Loads, Stable	Recurring imports and foreign exchange
Least Pollution and GHG Impact among Fossil fuels	
Moderate Water Requirements	
FURNACE OIL (Steam Plants)	
Fuel Supply Ease	Can be dirty depending on Sulfur level
Reliability	Generally most expensive among Fossil fuels(10-20 cents)
No Capacity Limitation	Price Risks
Base Load Operation and Constancy	Polluting and GHG
Moderate Investment	To be installed in no option case
	Recurring Heavy Imports and Foreign Exchange drain
Moderate Efficiency (38-40%)	Installed usually by Oil producing Countries or Peak Operations(HSD
COAL (Thar)	
Abundant in Pakistan ^ Elsewhere	Dirtiest Fuel, High GHG
Cheap and relatively stable(6-8 cents)	Location Issues
Moderate Investment(1.1-1.5 USD/MW)	Heavy Cooling Water Requirements ^ Limited availability in Thar
Employment creating	Transportation and Transmission Issues
High Capacity Factor (80%)	10,000 MW purported limit on Thar
Moderate Efficiency increasing (38-47%)	Power Concentration limitation
Base Load Operation, constant and Reliable	Import requirements dilute attractiveness
Energy Security	Solid Waste management Issues

NUCLEAR POWER	
Base Load, Constant Generation	Most Expensive(12 cents)
Cheap but imported Fuel	Most Capital Expensive(5 million USD/MW)
Reliability	Large Implementation Time(7-10 yrs)
No conventional Pollution and GHG	One Source Monopoly
	International Opposition
	Safety Risks
	Waste Management Issues
	Space Intensive, Siting Issues
	Coastal Sites in Balochistan
	Seismic and Tsunami Risks
	Fuel Supply Risks
HYDRO	
Clean and Renewable	No More Cheap(8-12 cents)
All Loads, Versatile	High Capital Investment(2.5 Million USD/MW and more)
Water Storage Possibilities	Large Lead Time(7-10 yrs)
No Pollution, NO GHG	Construction, Site and logistics Difficulties
Reliability	Localized
No fuel and import Liability	Low Capacity Factor (40-45%)
Energy Security	Displaces Populations
	NHP, Royalty Payment Requirements
	Dam Failure Risks
	Internal and International Political Issues

Source: Compiled by author

Table 4 Comparative parameters of Various Energy Sources

	CAPEX	COGE	Capacity Factor	Efficiency
	MnUSD/MW	Usc/kWh	%	%
Solar	1-1.2	5	17	
Wind	1.2-1.5	6-7	35	
Hydro	2.5+	8-12	45	
Gas(NGCC)	1.1	8-10	80	60
Oil-Steam	0.9	12-20	80	40-45
Coal	1.1-1.5	8-10	80	38-47
Nuclear	5	12+	90	38

Source: Compiled by author from various national and international resources

Renewables

• Wind and solar account for 48% of installed capacity and 34% of electricity generation world-wide by 2040. This is compared with just 12% and 5% today. Half of European electricity supply in 2040 comes from variable renewables, posing challenges for grid and generators. We anticipate renewable energy reaching 74% penetration in Germany, 38% in the U.S., 55% in China and 49% in India by 2040 as batteries and new sources of flexibility bolster the reach of renewables.

• The levellised cost of new electricity from solar PV drops by 66% by 2040. By then, a dollar will buy 2.3 times as much solar energy than it does today. The levellised cost of new electricity from onshore wind drops 47% by 2040, thanks to more efficient turbines and streamlined operating and maintenance procedures.

• Onshore wind costs fall fast, but offshore falls faster. We expect the levellised cost of offshore wind to decline 71% by 2040, helped by development experience, competition and reduced risk, and economies of scale resulting from larger projects and bigger turbines.

Coal

• Global coal-fired power generation peaks in 2026. Growth in coal demand is centered on Asia, but is offset by sharp declines in Europe and the U.S. Coal-fired generation in China

• Peak coal is in sight in Asia. Peak coal capacity occurs in 2024, and peak generation in 2028, as retirements begin to outpace new additions. By the mid-2020s, cheap wind and PV begin to undercut new coal on a levellised basis throughout the region, trimming average installations to just 9GW a year. Coal, however, remains the bedrock of the region's power supply, providing 34% of electricity in 2040 – a larger share than any other fuel.

• Coal consumption in China peaks in 2026, but at a level 20% higher than today. Nevertheless, China remains the world's largest coal consumer and emitter, with that fuel still accounting for 30% of the generation mix in 2040.

• India significantly expands its coal fleet over the next five years, adding over 40GW of new coal plants. Following that, we expect coal new build to slow but existing plant utilization to increase, pushing up coal consumption by around 3% per year through the 2020s. From 2030, solar begins to sideline coal in India, with the pace of PV additions more than doubling from the 2020s to the 2030s.

• Japan and South Korea shift from gas to coal, and then to solar. Gas generation declines in both countries as over 30GW of coal capacity is commissioned over the next decade – Japan and Korea are the only two members of the OECD to build significant volumes of new coal in our forecast.

Source: Bloomberg New Energy Finance
June 15, 2017

4. Achieving Universal Energy Access in Pakistan[1]

Access to clean, reliable and modern sources of energy is critical for achieving the targeted economic growth and development in a country. Vision 2025, formulated by the Ministry of Planning, Development and Reforms of Pakistan, recognized the importance of energy access; Pillar IV of Vision 2025 aims to ensure uninterrupted access to affordable and clean energy for all sections of the country's population.

Most rural households in Pakistan remain in a state of energy poverty. Without access to conventional energy sources like electricity and natural gas, people here, like in many corners of the globe, use a variety of non-conventional energy sources.

Access to Electricity

51 million people (27% of the population) in Pakistan still live in areas where electric grid is yet to reach. This means 8 million households (with an average household size of 6.35) in various parts of the country are living without access to electricity. A comparison of electrification rates across countries in the neighboring region shows that Pakistan is behind China, Sri Lanka and Nepal - all of which have electrification rates above Pakistan's. India also has a higher rate compared to Pakistan but owing to its large population, number of people without electricity in India (244 million) far exceeds the population in Pakistan without access to electric grid.

---

[1] This chapter is contributed by Saadia Qayyum, the editor of this book. Saadia Qayyum is an Energy Consultant and has worked for USAID, ADB, UNDP and World Bank. She has done her Master's in Public Policy from Harvard's Kennedy School of Government.

Table 5 Comparison of Electricity Access 2016

Region	Population without electricity millions	National electrification rate %	Urban electrification rate %	Rural electrification rate %
Bangladesh	60	62%	84%	51%
India	244	81%	96%	74%
China	0	100%	100%	100%
Nepal	7	76%	97%	72%
Pakistan	51	73%	90%	61%
Sri Lanka	0	99%	100%	98%
Other Asia	29	35%	66%	24%

Source: IEA Energy Database 2016

More than 32,000 villages in the country continue to remain without electricity grid forcing the residents to use traditional sources of energy, including firewood, kerosene and diesel, for meeting their lighting, heating and cooking needs. For most of these villages, sparsely distributed population and remote location has made expansion of grid financially unviable. Among the provinces, Sindh has the highest number of un-electrified villages, followed by Punjab, KPK and Balochistan. On the other hand, AJK and Gilgit-Baltistan compared to the rest of the country has more than 90% electrification.

An important thing to note is that connection to the grid does not equate to availability of electricity. Majority of the villages, which are officially listed as electrified, continue to experience 12-16 hours of blackouts thus being forced to spend more than half of the day without electricity. According to the Solar Consumer Perception study conducted by IFC Lighting Pakistan program, 71% of the country's population (144 million people) has either no access to electricity or experience more than 12 hours of blackouts. Thus the goal for the government is to achieve 100% electricity access in all the regions and to ensure 24 hours of uninterrupted supply of electricity to the customers.

Figure 4 Village Electrification in Distribution Companies

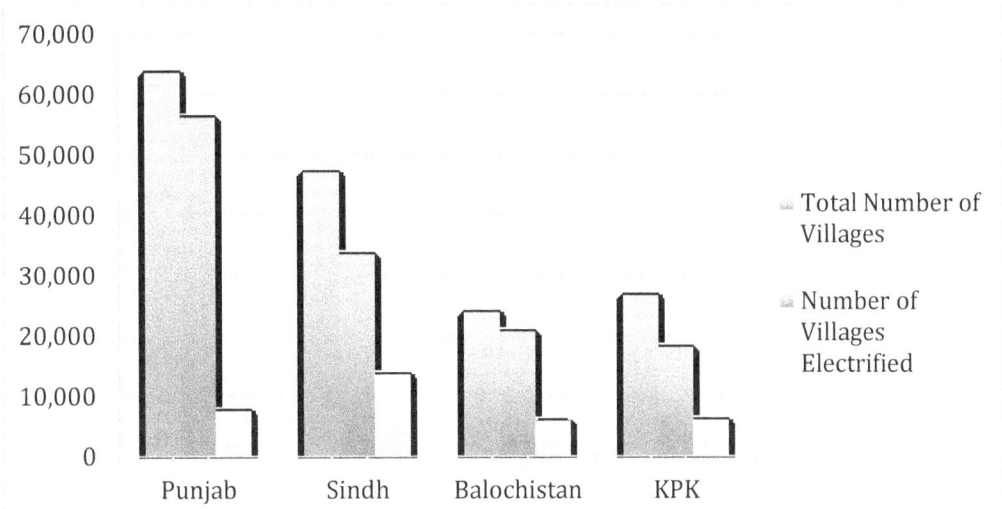

The off-grid energy sources that are used today are outdated and expensive such as lighting supplied by candles, kerosene lamps or battery-powered flashlights. All of these technologies are far costlier than grid electricity. These also pose risks in terms of damage to health, environment and safety. Although kerosene and candles are the most widely used form of lighting in off grid households, battery-powered torches are also used as often as gas-lights, generators and rechargeable lights across all provinces.

Distributed Generation in Off-Grid Areas

In areas such as these, it would make economic and financial sense to set up decentralized plants near the load centers directly supplying electricity to the consumers. With the advancement in technology and reduction in per unit price of solar and wind technologies, setting up off-grid plants will cost less than what would be required to expand electricity network to remote areas.

Governments, private developers, and NGOs throughout the world are increasingly pursuing microgrids to electrify communities that are unlikely to be served in the near or medium term by extension of central grid. These microgrids provide a range of services, from residential lighting alone to entertainment, refrigeration and productive commercial uses like milling.

In Pakistan, the penetration of microgrids, except in the northern part of Pakistan, remain low as business models for commercially viable operations of off-grid RE projects are not well established. Very few micro grid projects were found to be sustainable in the long run as they relied heavily on grants and subsidies from the government and/or donor agencies falling into problems of inadequate operation and maintenance and lack of cost recovery.

Notwithstanding these issues, increased deployment of microgrids in the country can play a pivotal role in powering un-served and underserved communities in remote locations where otherwise grid extension can be financially unviable.

Micro-grids employ various generation resources that include diesel, solar photovoltaic (PV), micro-hydro, and biomass gasification, as well as hybrid technologies such as wind-diesel and PV-diesel. In provinces such as Sindh and Balochistan, where solar irradiance is the highest and pockets of high speed winds are found, solar and wind hybrid microgrids can be set up to electrify remote and dispersed communities. Similarly in the north, where there is abundance of hydel potential, micro-hydels can serve the same purpose.

Just as villages that never saw the expensive telephone network jumped straight to mobile phones, improving economics of decentralized generation technologies offer the chance for them to leapfrog the electric grid. Even though the amount of power made available through these off-grid systems is lower and quite basic, they allow the beneficiaries to shift from traditional to modern lighting systems. This transition can stimulate employment and development opportunities in the region as community members can benefit from the power provided to schools, hospitals, water-supply systems and communication facilities.

Access to Modern Fuels

Traditional fuels like firewood, dung and crop residues currently contribute a major share in meeting the everyday energy requirements of rural and low-income urban households in Pakistan. According to World Energy Outlook 2014 data, 62% of the population (112 million people) continues to rely on use of biomass.

A large proportion (61%) of Pakistan's population still lives in villages (World Bank Data, 2015)[2]. As a result, this component is much higher (90% compared to 50% in urban

---

[2] http://data.worldbank.org/indicator/SP.RUR.TOTL.ZS

areas) in rural areas with dominant use of firewood for cooking and heating purposes[3]. As shown in the figure below, almost 60% of the country's population uses wood for cooking, followed by gas (32%) and dung (7%).

Figure 5 Percentage of Population using Traditional Sources of Fuel

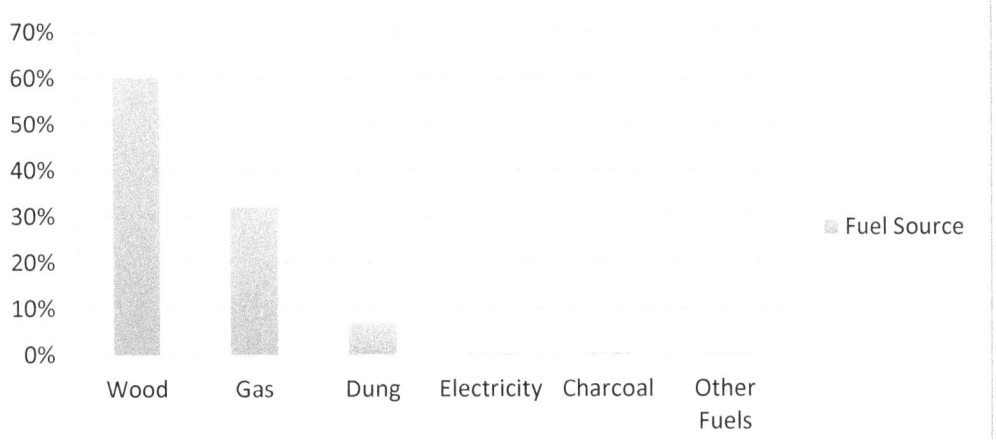

A village woman normally spends up to two hours a day making dung-cakes or collecting firewood depending on size of her family and amount of cooking fuel required. Amount of energy required for cooking varies with the type of food, the fuel and stove used and the specific cooking practices of a household.

Young children are often carried by mothers or kept in the kitchen area during cooking exposing them to high levels of smoke. Because women do most of the cooking and spend more time indoors, they are exposed more to pollutants and are believed to have greater adverse health impacts. In general, rural women and children are malnourished and the impact of indoor air pollution on them is likely to be much stronger. The table below shows the health effects of air pollution caused by use of traditional fuel.

---

[3] WHO, Situation Analysis of Household Energy Use and Indoor Air Pollution in Pakistan < http://apps.who.int/iris/bitstream/10665/69705/1/WHO_FCH_CAH_05.06.pdf>

Table 6 Health Effects of Hazardous Air Pollution

Number of People Affected by Hazardous Air Pollution (HAP)	111,079,269
Number of People Households Affected by HAP	16,335,187
Number of deaths per year by HAP	114,806
Number of child deaths per year by HAP	33,673

Source: Global Alliance for Clean Cook Stoves, Pakistan Profile

Indoor air pollution from cooking and heating with traditional fuels has been designated by the World Bank as one of the four most critical environmental problems in the developing countries. In Pakistan, 111 million people are exposed to the negative health effects of air pollution emitted by use traditional fuels.

Access to gas in the country is as low as 25%. Due to the capital cost of laying gas pipelines in the mountainous terrain, both AJK and Gilgit-Baltistan do not yet have a piped gas network and have to rely on either expensive liquid petroleum gas (LPG) cylinders (transported from down-country) or burning of firewood for fulfilling their heating and cooking requirements. This poses not only health hazards for the residents but also environmental problems in the region. The direct effect of collection of firewood and construction timber without forest management or adequate re-plantation has caused the tree cover in the mountain areas to become less dense.

Transitioning to Modern Fuels

Biogas

There are good prospects for using biogas energy in rural areas of Pakistan through a network of community biogas plants. The amount of dung-waste is enough to produce about 12 million cubic meters of biogas per day that could suffice to meet energy requirement of 28 million rural people, in addition to production of 21 million tonnes of bio-fertilizer per year. It is reported that presently there are 5,357 biogas units installed in the country. The unit sizes are 3–15 m3 /day. The estimated countrywide biogas potential is 12–16 million m3 /day.

Recently, PCRET has installed 4000 biogas plants throughout Pakistan on cost-sharing basis, where 50% cost is to be borne by the beneficiary. Apart from these, three community size biogas plants have been installed in rural areas of Islamabad, which are meeting domestic fuel needs of 20 houses.

The more improved alternative available to rural villages is biogas plant extensively applied in Europe and India. In India more than two million biogas plants are in use and each year 200,000 families replace their traditional fireplace with a biogas plant for cooking and heat. Similar equipment has been used for gas production with domestic wastes.

Fuel-efficient stoves

Improvements in the efficiency of the conventional cook stoves can substantially reduce the consumption of wood, thereby mitigating harmful air pollution, saving household expenditure and slowing down deforestation.

Technical advances in fuel-efficient cook stoves, therefore, are crucial, for the rural as well as urban population heavily depend primarily on biomass. PCRET has installed 60,000 energy-conserving, improved cooking stoves, so far, all over the country, which are 12–28% more efficient. Whether these cook stoves are still functional and in use is not known.

SAARC Energy Center (SEC) initiated a study on Improved Cooking Stoves (ICS) in South Asia in 2010 with an aim to cover the necessity of ICS in view of biomass dependence in the SAARC region and impact of cooking with traditional stoves. Subsequently, the program to develop, design and manufacture Improved Cooking Stoves (ICS) was initiated by SEC and it was successfully completed in 2012. SEC has developed three different designs of ICS for different climatic regions of SAARC Member States. These ICS were deployed in the field for trials in 2013 in different climatic regions of Pakistan simulating the climatic regions of SAARC Member States.

In one of the improved stove programs, a Lahore-based NGO with funding from UNDP has introduced a fuel-efficient cooking stove, to relieve pressure for fuel wood in the Changa Manga forest and to improve the health of village women. The women who use them make the stoves, out of mud and straw. These consume less than half the fuel wood.

5. Are Energy Prices High in Pakistan

Petrol Prices controversy

Petrol is a commonly used term implying both petrol and diesel and we will discuss both in this space. It is a general impression and complaint in Pakistan these days that the petrol prices have not been brought down in Pakistan sufficiently in the wake of drastic reduction in international oil prices. It is alleged that elsewhere and in the region, governments have passed on the benefit to people and consumers due to which inflation has been reduced there and competitiveness increased. It is also alleged that one of the reason of rising cost of production and falling exports is the allegedly wrong policy of the government in this respect.

Let me put the facts right first. In the adjoining table 7, we provide a comparison of the petrol prices of Pakistan vis-à-vis other selected countries including India and Bangladesh. These are as latest prices (22 May, 2

017) prevailing in the respective countries as we could get hold of at the time of this writing. In India, there is regional pricing of petroleum. We have taken the retail prices of petrol in Mumbai where significant petrol consumption is there and market operates relatively well.

It can be seen that both diesel and as well as petrol prices in US dollars in India are higher than in Pakistan. In fact reverse has happened. In India, diesel prices used to be lower than in Pakistan and petrol prices used to be and are higher than in Pakistan. Now, in case of both, petrol and diesel prices in Pakistan are lower, and quite significantly at that, as can be seen readily in the table. It is ironic that not only TV anchors and ordinary discussants have taken this wrong view of higher petrol and diesel prices in Pakistan, but some noted economists who have also taken this position, reflecting sheer venom, as I cannot allege that they are not knowledgeable. Or it has become a national pat time to comment without checking the fact.

Table 7 Gasoline and Diesel Prices in Selected Countries (May 22-2017)

Country	Gasoline	Diesel
	USD/Liter	USD/Liter
USA	0.7	0.66
Pakistan	0.71	0.79
China	0.96	0.85
Bangladesh	1.05	0.77
India	1.08	0.9
Turkey	1.44	1.25
Singapore	1.44	1.01
Germany	1.47	1.23
France	1.49	1.32
UK	1.5	1.51
Italy	1.67	1.5
Norway	1.83	1.68

Source: www. GlobalPetrolPrices.com

A possible reason for this misconception appears to be the increase in taxation in relative terms. GOP claims that it has kept the taxation constant in absolute terms, which due to reduction in oil prices, shows itself as high in relative terms. Otherwise, who can deny that we have paid as much as Rs 125/- per liter in the past and now the prices (of Petrol) are in the range of Rs 70-80 per Liter?

Petroleum has traditionally attracted high taxation in many countries. In Europe, petroleum prices are the highest in the world, reaching around Rs.200/- per liter in some countries due to more than 100 % taxation in these countries. Petrol, if not, diesel has been traditionally considered as a luxury item, although no more. It is a necessity. Governments in Europe have become used to this policy. Old habits seldom die and petroleum provides a significant revenue to governments. Petroleum taxation is also considered as a user charge for roads and transportation infrastructure and in case of Pakistan, taxation of petroleum pays for subsidies in electrical sector. If you are not happy with petroleum taxation, then be ready to pay still higher electricity bills. Net effect would be the same. Overall, it appears that energy sector as a whole has zero taxation in indirect terms (excluding income tax).

To put it in perspective, there are four regions/groups, following four different pricing policies as has been described in the following:

- High taxation and high petroleum prices such as in Europe generally except in Spain and a few others
- Significant taxation and moderately high prices such as in most developing countries including Pakistan, India, African countries and others
- Light taxation and low prices e.g. the U.S. and others
- Subsidy pricing countries, mostly oil producers and exporters, like Iran, Saudi Arabia, UAE, Venezuela, where even production cost is not recovered. Only lately, these countries have started increasing prices due to shortfall in oil revenues.

Concluding, there are long-term policies and pricing trends in countries and politics and changes thereof rarely affect these in a significant way, as economies' structure is built around these issues. It is high time that partisan and incorrect views are abandoned by our electronic media, politicians and academicians. We will take up the abnormally high rate of growth in Petrol consumption and generally the demand issue in later discussions.

Electricity and Gas Tariff

We have discussed Petroleum products prices in the foregoing, wherein we have concluded that Petroleum prices are the lowest in the region these days, contrary to the popular criticism. In earlier days, petrol prices used to be cheaper in Pakistan and diesel used to be cheaper in India. No more, as prices of both the products now are lower in Pakistan than India and even Bangladesh. Let us look at the gas and electricity prices. Table 8 provides regional data on electricity tariff.

Bangladesh has the lowest electrical tariff as can be readily observed in the table above. A general statement about tariff in India is difficult to make. Electrical tariff in India is different in every state, unlike in Pakistan where it is almost a uniform tariff. In Pakistan, subsidies are given to the small consumer category and probably to a limited insignificant impact end-users. All others pay its full price. We have chosen TATA's rates in Maharashtra. There are two other companies, Reliance and Best active in the electricity distribution there. Maharashtra, Kerala, Karnataka, and Andhra Pradesh etc. are high tariff states, while Haryana, Punjab and Gujarat are lower tariff states. TATA and KESC tariff for domestic consumer is almost comparable. In case of small industry (5 kW), the tariff is also the same, almost. It is in case of large industry, TATA (Maharashtra) tariff is Rs.4.00 per unit lesser than it is in Pakistan. Perhaps, it was this reason that Rs. 3.00 per unit reduction in tariff was recently awarded to the textile sector. It seems to be the remission in GST which comes out to be that much.

Table 8 Comparative Electricity Tariffs in India and Pakistan (USc per kWh)

Tariff Category	TATA-Bombay*	Pakistan	Bangladesh
Domestic			
0-100 units	5.328	5.79	4.56
301-500 units	16.352	16	6.432
500+	20.352	18	11.976
Industrial			
20 KW	14.144	14.5	9.192
20 KW +	14.752	18.77	8.988

*These tariffs also approximate those in N.D., Karnataka, Kerala, Andhra Pradesh, etc. Elsewhere in India, e.g., Punjab, Haryana, Gujarat, Rajasthan etc., tariff is lower

1 USD=IRs 62.50=Btk80.60=Pk.Rs100

Source: TATA,DECO and KESC websites

MF Almanac says that there should be no subsidies, as it is argued, that it leads to wastage and overconsumption and misallocation of resources. It may be at-least partly true. Generally, industry in Pakistan is poor in conservation and energy efficiency. Lower than actual prices may make them still more lax in conservation. And then where do you get the money from, when you sell a commodity (electricity) at less than the cost. You would divert it from other less powerful and deprived sectors such as education and health sectors.

The way out is to reduce the cost of generation and reduce T&D losses. It is hoped that substitution of Furnace Oil by cheaper coal, RLNG (NGCC), Hydro and Solar would result in reduction in cost of generation. It should start happening from 2018, when new power plants come on stream. T&D loss reduction is a mid-to-long term issue, partly related to socio-political conditions, apart from technical issues.

In the short run, GST provides some handle on tariff reduction in priority sector. Malaysia levies GST of 6% on industry and exempts small domestic users (up to 300 units) from GST. In Singapore, GST on electricity is 7% and in Australia it is 10%.

Higher Energy Prices in OECD Countries

In Germany, domestic users' electrical tariff is at its highest, amidst news of cheaper renewable energy. It is true that that the tariff of new renewable energy plants is lower, but Germany's tariff system is a victim of earlier subsidies announcements, which have along life-cycle. In the U.S., electricity is the cheapest amongst the selected group of countries; only 12.40 cents as opposed to 34.7 cents of Germany. Needless to indicate that European tariff is quite higher than in South Asia.

However, in Industrial sector, the story is different. As usual, the US tariff is the lowest at 9.48 cents. It is, however, not very different than average industrial tariff in Europe; in France and Spain it is quite close at 10.4-12.04 cents and in the UK, Germany and Italy, it is around 15-18 cents. All states in the US have different tariff. There is a difference of 18 cents in the industrial and domestic tariff in case of Germany. Industrial tariff in all Europe is lower that domestic. When you confront Europeans, they mind it and yet are not able to give a clear explanation. They hide behind higher service charges in case of domestic users which could be only partly true. Why in the US tariff system, domestic and industrial electricity or even gas tariff much closer to each other than is the case in Europe. Is the level of service and care lesser that it is in the U.S. utilities?

Due to the lowest well-head gas prices and the cheaper Shale gas, the Gas tariff in the U.S. is comparatively low: Pk.Rs 990 (USD 9.9) per MMBtu for Domestic users and Pk.Rs. 402(USD 4.02) per unit for industrial users. In fact in all the selected countries, the gas tariff is almost uniform and comparable; around 11-13 USD per MMBtu for industry and 23-25 USD for domestic users. In Spain and Italy, it is higher than average. In Pakistan, current higher tariff for gas is 6 USD per unit, same for industry and high demand domestic consumers. RLNG users get the gas at 8-10 USD per MMBtu. Generalization is risky and can be misleading, especially in case of federal countries like U.S., where different economic policies and resource endowment leads to different prices. Thus we conclude that gas prices are also low in Pakistan.

Table 9 Gas & Electricity Prices for Household and Industry in Europe (2016-2017)

Country	Electricity		Gas	
	Household	Industry	Household	Industry
	Pk.Rs/kWh	Pk.Rs/kWh	Pk.Rs/MMBtu	Pk.Rs/MMBtu
France	19.89	10.413	2503	1269
Germany	34.749	17.433	2332	1303
Italy	27.378	18.252	3120	1097
Spain	27.729	12.051	3189	1097
UK	23.517	14.976	2297	1200
USA	12.4	9.48	990	402
Pakistan	16	18	600	600

6. Energy Prices and Economics in Pakistan

There is a wide difference in costs and prices of various energy sources; coal vs oil vs gas etc. There are factors other than taxation, which are responsible for these variations. We will examine here the cost structure of various energy sources that would enable us to develop an understanding and appreciation of the differences involved. Competitive factors cause a difference among cost and prices. We have examined this issue in another chapter e.g., the case of oil where large margins have enabled profits even when sales price has gone down by more than 50%.This means monopolistic influences on oil market which collapsed with contraction of demand. There is a controversy on Thar Coal and Energy tariff, the latter we have compared with imported coal tariff and that prevailing in India on Lignite and others. We have also provided comparison between RLNG and Coal based energy costs and tariff.

Comparison of Fuel Costs

If we compare the NEPRA monthly data on fuel cost of Jan 2017, the latest available at the time of this writing, RFO electricity variable fixed cost is Rs.8.8329 per kWh vs. Rs 7.3641 per kWh of RLNG based electricity; RLG being 83.3% of Furnace Oil Electricity. This data is on the existing low efficiency steam power plants. Thus RLNG based electricity is costing Rs 7.36 per KWh as opposed to the estimate cost of Rs.6.00/kWh estimated or even slightly less for new high efficiency Combined Cycle power plants being installed at Bhikki and others.

Oil prices came down to as low as 30.80 USD per barrel in January 2016 but have recovered gradually since then and were 51.97 USD in last March after achieving a peak price of 55.48 USD a month earlier. Oil prices, according to many expert agencies, are expected to remain at about this level in medium term.

Table 10 Comparative Fuel Cost in Electricity Production (Rs/kWh)

	Jul-16	Jan-17
Coal		4.4998
HSD	12.7632	13.7157
RFO	7.5032	8.8329
Gas	5.1186	5.1702
RLNG	5.3452	7.3641
Nuclear	1.1603	0.81
Mixed	7.7288	6.5164
Bagasse	6.0154	4.9014
Total		6.3956
CPP and adjustments		1.5427
Transmission losses		0.3361
Net DISCO Price		8.2455

Source: NEPRA

It is very difficult to make price forecasts, as these often turn out to be incorrect widely. Take the example of coal; in early 2016, many people said that coal is going down to its grave at 50 USD per tonne of prices. Coal recovered in next six months and the same people now say that coal is rising from grave to be one of the hottest commodities. Coal prices were as high as 80.62 USD per ton in the last month. Natural gas prices in Europe have almost remained static around 5.35 USD per MMBtu. LNG prices (in Japan) remained rather stable around 7-7.5 USD per MMBtu after dipping to 5.86 USD in May 2016.

Times are changing, especially in LNG sector. There was a time when LNG to Oil Price ratio used to be above 80%, based on which Iran Pakistan pipeline gas prices were finalized. Now the lowest price gives us a corresponding ratio of 64.51%, which means that in Btu terms, LNG prices are 64.51% of the oil prices. This is the reason that Minister PNR is quite upbeat about it. But things change as we have mentioned elsewhere. Supply and demand and the perception of it among players establish prices, which is often not predictable. Coal was going to its grave, according to many experts at 40-50 USD per ton only a year earlier, it is now selling at double of those prices. If LNG prices remain as it is, LNG can play a major role in our energy supplies; otherwise we will have to hedge our risks.

Thar Coal vs. Imported Coal

The Cost of Generation (COGE) of imported coal electricity is estimated as Rs.8.0139 per kWh while that of Thar coal Rs.8.50. There are a lot of components that have to be kept in view and these have been provided in Table 11. The first question that emerges is that why Thar coal electricity is expensive, even though, it is local and no import costs are involved? Normally, imported coals have 40-50% of cost going to transport and handling. The fuel component of Thar coal is in fact costing lesser (Rs 3.54/kWh) than the imported coal (Rs.4.92/kWh). These are reference prices assumed at the time when tariff was being calculated. Imported coal at that time was USD 90 per ton FOB, which is now 65-69 USD per ton FOB.

In Europe, lignite production cost and prices are around 20 USD per ton mine-mouth. Lignite has half the Calorific value (energy content) than that of normal sub-bituminous coal that is used in power plants and is being imported in Pakistan as well. Leaving aside the question of Thar coal prices i.e. whether these should have been even lower, let us try to explore the answer as to why Thar coal electricity is expensive? The only factor that remains is of thermal efficiency and CAPEX and Profits. Thermal Efficiency of Lignite power plants (37% presently of Engro) is normally slightly lower than that of imported/subbituminous (at 40%), although, these days, thermal efficiencies have gone to a level of 45% for Super-critical and Ultra-super-critical power plants, which normally are more expensive in CAPEX. Apparently, CAPEX of Engro-Thar is higher (1.50 Mn. USD per MW), as compared to imported ones (1.35 Mn USD/MW). However, if jetty cost is added for imported coal projects, the difference goes away. In imported coal projects, jetty costs have been subsumed in fuel running cost. The real culprit is the profits or returns allowed, which has been widely criticized, for right and wrong reasons. The right reason for criticism is that the IRR allowed is too high at 20% as opposed to 15% generally and 17% in the case of imported coal. 20% IRR is quite high as incentives of 1% are considered enough and significant internationally. Resultantly, the fixed cost component alone of Thar power projects is Rs.5.0 per unit, while elsewhere the total cost including fuel is that much. This is discussed in detail in later chapters.

Apart from increasing the electricity cost, such atrociously high returns are bad for Thar coal itself rendering it uncompetitive. It is already so, if current lower international coal costs are considered; 20 USD per ton for Lignite and 65-69 USD per ton of sub-bituminous. Generalization is difficult, however, normal coal electricity prices in the world are 5 cents or lesser which includes 40% corporate tax (as opposed to 7.5% tax on

Table 11 Comparative Coal Prices

	Units	Thar-Engro	India lignite	India Sub bituminous	Ref Tariff imported	Current
FOB	USD/ton	50.69	34.68	30	90	69
Transport cost	USD/ton				20	20
Landed Cost C&F	USD/ton	50.69	34.68	30	110	89
other handling	USD/ton				20	20
Total at plant	USD/ton	50.69	34.68	30	130	109
CV	Btu/lb	5000	5000	6840	10253	10253
CV	MMBtu/ton	11	11	15.048	22.5566	22.5566
Coal price	USD/MMBtu	4.6082	3.1527	1.9936	5.7633	4.8323

Source(s): NEPRA, CERC India, GEMC India

dividend here) and environmental protection costs, both in terms of CAPEX and OPEX. We are only installing Electrostatic Precipitators (ESPs[4]) to control particulate matter. I have been arguing against unusually high returns that have been allowed (let us not hold NEPRA alone to be responsible in it; it is the system).The investors earn in many other ways such as savings or padding in CAPEX, saving in fuel and operating costs. A usual saying in the market is that investors do not really put in their own money. If one examines the enclosed table that is provided, and if optimization is done and unnecessary provisions (high profits, high risk margins of 3% in local interest and 4.5% in foreign loans, despite Sinosure risk coverage of 7%, high insurance rates etc.), a considerable saving can be made and coal electricity can be brought down at a reasonable tariff of 5-6 cents. Let us try it, now that we can take some risks, as 10,000 MW is coming in already.

---

[4] An electrostatic precipitator (ESP) is a filtration device that removes fine particles, like dust and smoke, from a flowing gas using the force of an induced electrostatic charge minimally impeding the flow of gases through the unit

Table 12: Comparative Tariff: Thar vs. Imported Coal

	Units	Imported Coal	Thar Coal
Capacity	MW	1,099	330
CAPEX	Million	1,483	498
CAPEX per MW	Million USD/MW	1.350	1.508
Thermal Efficiency	%	40	37
Tariff	Usc/kWh	8.014	8.502
Energy Cost	Usc/kWh	4.92	3.54
CPP	Usc/kWh	3.10	4.96
ROE	%	24.5	30.65
Coal price	USD/ton	129	50.69
Coal price	USD per MMBtu	5.76	3.84
CV	Btu/lb	10,253	6,000
Capacity factor	%	82	85

Source: NEPRA Tariff Determinations

RLNG vs. Imported Coal

Lower gas and RLNG prices and a higher thermal efficiency (57% unprecedented) and ill lower CAPEX has brought RLNG into quite a competitive situation which has impelled some people in the O&G sector to conclude that RLNG is the solution to Pakistan's problem, a debatable hypothesis, that we have dealt with separately, and will come back on it at a later opportunity. Suffice to say here that, the grapevine of lower LNG prices may not continue indefinitely into the future. What happens if the prices are doubled or tripled as earlier? It has happened in the past and may also happen in the future. LNG prices have already started going up. Price risks have to be balanced in the form of a composite energy mix. Local coal or renewable energy would not be susceptible to international price risks. However, internationally, one would face increasing difficulty in financing coal power plants due to mounting opposition to coal due to climate change issue. All energy sources have risks of some sort. There is no panacea.

In Table 13, we provide data on RLNG based combined cycle power plants. As one would readily note that COGE of RLNG electricity is comparable with coal, assuming an RLNG price of 10 USD per MMBtu and a reasonable long term capacity factor (utilization) of 80%. The energy component of the tariff is about Rs.6.00/kWh. However, if cost rationalization is done as discussed earlier, some cost reduction is possible here also, although the tariff basis are more realistic and lower such as the allowed IRR on Equity of 15%. CAPEX of 1.046 Million USD per MW is also reasonable, lower than earlier ones, and lowest of all.

Table 13 Cost of Generation (COGE) of RLNG NGCC vs. Imported Coal

	Unit	RLNG	Imported coal
Plant Capacity	MW	800	1099
CAPEX	Mn.USD	863	1483.7
Unit CAPEX	MnUSD/MW	1.046	1.35
Interest Rate Foreign	%	4.7556	4.7556
Interest rates local	%	12.75	
IRR on Equity	%	15	17
RLNG Price	USD/MMBtu	10	5.7633
EPP	Rs/kwh	6.1204	4.9161
CPP (80%)	Rs/kWh	2.1779	3.0978
TPP	Rs/kwh	**8.2983**	**8.0139**
Thermal Efficiency	%	57	40
Construction Period	months	28	40

Source: NEPRA Tariff Determinations

Attempts were made by the rentiers to charge higher overheads such as 4% similar to service charges for PSO and the likes, to add the cream for themselves and save the same from the people. Fortunately, Minister of Petroleum was vigilant and mindful and was enthusiastic for his pet project and OGRA did its job well as well and reasonable charges were levied. Of course much could not have been done for regasification charges of 66 cents for an already implemented project. New RLNG terminal is charging 41 cents and still newer may further reduce this cost as alternative technical and financing approaches are adopted and as have been suggested elsewhere.

LNG vs Local natural gas: the role of competition and incentives

Pakistan and Qatar LNG agreement attracted quite some controversy, which died down appreciably after the MPNR started international bidding. Actual LNG market has evolved over the past year and has kind of materialized. Minister PNR thought that he clenched a very good deal from Qatar. Reportedly, prices are still under negotiation with Qatar in the light of international tender results, as we shall discuss. International tenders had been invited for the supply of LNG. Results of the lowest bids have been provided in Table 14. Guvnor offered the best price at 11.62% of Brent, which is lowest price offered for 5 years of short term contract. The next best is ENI with 12.29 % for a contract period of 15 years. These days longer term contracts are more expensive due to risks of price and supply.

Table 14 LNG Supply Contracts and Prices

	LNG Price	LNG Price	Oil Price	LNG-% of Oil price
		USD/MMBtu	USD/MMBtu	%
Pakistan-Qatar	13.37	6.685	9.007	74.22
Guvnor	11.62	5.81	9.007	64.51
ENI	12.29	6.145	9.007	68.23
MMBtu in one barrel	5.5513			
Assumed Oil Price/bbl.	50			

Source: MPNR, The News et al

International tenders have once again demonstrated the merits of open competition. One can discover the true price only through competition. Even suppliers themselves learn the price through competition. We thought that G-to-G contract may give us a better price but It did not, although the Minister defends it by explaining the contract differences; Qatar contract is not a take or pay contract. This can be a significant concession, if true and implemented accordingly. There were people at Planning Commission who argued for international tender in the first place. Fortunately, sense prevailed ultimately and almost best prices were obtained, for the time being.

The Minister insists that he got a better deal than India, who got 20 % higher prices from Qatar. Contracts are not always comparable and contract clauses are often confidential, at least not available on the internet. Whatever little we know of India's LNG contract with Qatar, there is an additive term in the Brent Formula, which makes Indian contract more expensive, even if Brent number is the same.

Times are changing, especially in LNG sector. There was a time when LNG to Oil Price ratio used to be above 80%, based on which Iran Pakistan pipeline gas prices were finalized. Now the lowest price gives us a corresponding ratio of 64.51%, which means that in Btu terms, LNG prices are 64.51% of the oil prices. This is the reason that Minister PNR is quite upbeat about it. But things change as we have mentioned elsewhere. Supply and demand and the perception of it among players establish prices, which is often not predictable. Coal was going to its grave, according to many experts at 40-50 USD per ton only a year earlier. It is now selling at double of those prices.

Yet it is not enough to get the lowest and the best prices in LNG procurement, as there are a lot of add-on costs that may add 33-50% to the basic cost involving many intermediaries such as regasification terminals, traders, and transmission and distribution companies. Fortunately, OGRA has managed to prevail and has approved reasonable add-on components.

Regasification costs of the first terminal at Port Qasim is rather excessive at 66 US cents per unit (MMBtu). The second one is only after a year's gap and is costing 42 cents, even as a result of an imperfect competition (not participated widely). These costs are based on excessive leasing rates. If alternative debt financing is involved at 1-2% of ADB/World Bank interest rates vs 15-20% leasing rates, these costs can be brought down to around 22-25 cents. Even local banks can finance these projects of 250 million USD CAPEX at rates of 6-7 % these days. Reportedly, MPNR is actively considering alternative technical and financial approaches in its new projects. It is expected that eventually, the difference between long term and short term contracts prices would be either eliminated or reduced appreciably and the next round of bids may yield better results.

Local Gas Prices

Our local well-head prices are around 3.0 USD per MMBtu (Table 5), as against landed LNG price of around 6.0 USD per MMBtu. Our known reserves are falling down. New FDI in Oil and gas sector is not forthcoming. At low prices, investment in parent countries is stagnant, not to talk of here. FDI is not just money; it means technology of finding and developing oil and gas. Earlier the local well-head prices were at 5-6 USD per MMBtu when Oil prices were high, almost double that of today. An important issue at this stage is whether we should keep importing LNG at double the rate of local gas, or should we increase the price of local gas i.e. of new discoveries and not the existing ones, which may incentivize our local production leading into new discoveries and development of new fields. It has always been a question, which could not be settled.

Table 15 Natural Gas Prices in Pakistan (USD/MMBtu)

Well-head Prices	
Badin I,II,III	2.5277
Chachar	1.2884
Dakhini(Rs per MMBtu)	154.5
Dhodhak(Rs per MMBtu)	308.7
Kandanwari	3.336-3.9587
Mari	0.8405
Miano	2.6238
Neshpa	2.715
Sawan	2.6238
Sui(Rs. Per MMBtu)	132
Uch	3.9223
Zamzama	2.5339-2.6239
Average Price(Rs per MMBtu)	341.25
Gas Prices for IPP	340.52
for Industries	340.52
for Commercial	444
for Residential	74-370

Source: Energy Yearbook-2015

Local Well-head prices(for new discoveries) may have to be linked with LNG and brought at par with landed prices .There can be floor and roof price provisions. And the high prices may be for the first five discoveries made in the first five years. Thereafter, one could readjust the incentives keeping in view the prices. However, this may not be the right time just before elections. Consultations may be initiated at this time and policy papers and proposals can be prepared.

7. Where will Oil and Energy Prices go?

If you are a consumer or an importer, you would like that oil and energy prices are as low as these are for the last couple of years, and continue to do so for as long as possible. If you are a producer or an exporter, you would like the reverse to happen. In Pakistan, purchasing power is so low that even these low prices are not affordable but the governments want to extract as much from oil as it can. In this article, we would like to apprise the reader of the possible oil and energy prices scenario in a context of the next two decades and the behavior of factors that go into. A tall order for the space that is available.

Oil and energy in general play a major role in our lives and economies - never before this role was as paramount as it is today. People cooked their food on wood and shrubs, as they still do in most of our rural areas even today, the tradable energy is what is new and that is why the subject of energy prices is so important and some sense of what these would be in future (for those who think of the future).

Oil's dominance can be measured by the fact that seven of the largest ten multinationals are oil companies, Royal Dutch Shell being the largest at 275 Billion USD in sales and revenue. The two super powers are among the largest oil producers, and one of them (US) is the largest importer and consumer of oil, consuming one-fifth of the world production.

Oil, gas, coal and electricity and to some extent renewable energy are the major elements in the energy mix of today. Prices are generally interdependent (one affects the other due to substitution effects and others), Oil prices being the most powerful driver and thus would dominate our discussion here.

Oil prices skyrocketed to USD 145/b in 2008. It leveled out to around $100/b in 2014. It plummeted to a 13-year low in January at USD 28 then doubled to current levels of around 50 USD. If shale oil producers go out of business, and Iran doesn't produce what it says it could, prices could return to their historical levels of $70 - $100 a barrel. OPEC is counting on it. So what would be the prices in 2020, 2030 like?

It is easier and plausible to talk of energy prices in long term rather than short term, which is based on pure speculations concerning almost everything. Consumers and producers behavior's and aspirations and determine the balance which determines prices. And within consumers and producers are individuals, companies and governments that may have congruent or contradictory requirements and attitudes.

Oil demand, as predicted by many credible international institutions, will continue to grow at least till 2035 and the supply would catch up accordingly. Oil supply is constantly increasing since 2011 and will continue to do so. Shale oil and gas are a major threat to OPEC producers and have brought the prices to the level at which these are today. Otherwise, the prices would have been in the range of USD 70-100 per barrel. Higher prices may not always be in producers' interest. Demand may go down, the biggest weapon in the hand of consumer to fight prices is to reduce or avoid consumption. It has happened many times in the past. That is why Saudis do not want to press for excessive prices that may boomerang to destructively low prices.

Cost of production does have a basic role in prices in the long run, although in the short run the role of demand and supply balance plays a higher role. In the long run, prices would asymptote (approximate) with the cost of production. But the cost of production in most countries is USD 20-25 per barrel. Why aren't producers satisfied at the current prices then? There is an interesting feature or concept in prices that is of Break-Even prices.

Lightly speaking, I would term break-even prices to be the feel-happy prices. At break-even prices (BEP), the producer meets his cash flow requirements and is able to cover all its extra cost and overheads over and above the direct production cost (the 20-25 USD per barrel). Libya would require the oil price to be 200 USD per barrel to meet its budgetary requirement and thus it would like to have high oil prices. Saudis and Russians say that they would require 100 USD per barrel to balance their budgets. They may not get it all, but this requirement would certainly influence the prices. The producer-consumer USA brings its shale resources into the market and Europeans bring in renewable energy and conservation to balance.

Then there are BEP's of major oil companies such as Shell, Exxon and others. BEPs of large companies used to be around 75-80 USD per barrel in just 2014 as opposed to the production cost of 20-25 USD per barrel that we have been talking throughout. They have cut corners and have managed to bring it down to a level of around 40 USD .No company would be able to borrow funds, if it is not able to prove that it would have its long term Break Even Price under USD 40 per barrel.

With this condensed yet elaborate perspective, let us turn to the forecast of EIA, one of the leading and credible institutions, less influenced by commercial interests and others. They have predicted a price of 92 USD per barrel in 2025 at constant prices of 2015 (the uninitiated may take it as prices excluding general inflation) and nominal prices of 112 USD per barrel (including general inflation). This is their reference or base case forecast; the other two are of either high prices and of low prices. At constant prices, in the base case scenario, oil prices are predicted to grow at a rate of 3.9% p.a. in constant terms and at 6.1% p.a., a bit on the higher side for my understanding. In the next 10 years, my assessment is that prices should remain under USD 100 with a bottom of 70 USD.

Oil producers do not make a difference in oil or gas in terms of their production cost assessments. Gas prices corresponding to the 20-25 USD per barrel, are 3.63-4.54 USD per MMBtu. Gas prices have varied widely in the three important market regions: the U.S., Europe and Asia and the difference persists even now largely due to low shale prices and high supply in the U.S. LNG has brought tradability and thus some uniformity in prices. Earlier LNG prices used to be 16 USD per barrel in Asia and 8 USD in Europe. I never quite liked or understood it and wonder as to why Japan tolerated it. Partly, transportation costs explain the difference. Natural Gas prices predictions as done by EIA are for the US itself. US natural prices have been predicted to be in the range of 5-6 USD per MMBtu for 2025, as opposed to 2.62 USD currently. One may add USD 3.00 to cover liquefaction and transportation cost to Asia. U.S. prices will have a major effect on gas prices elsewhere in future due to LNG factor and its tradability. NG prices in the US have been predicted to grow at a rate of 2.5% in constant prices and 4.77 % p.a. in current price terms, a substantial rate. It may be a sharp reminder to those who are arguing for a LNG-preponderant system in Pakistan.

Coal at mine-mouth is being sold at 1.69 USD per MMBtu or 30 USD per tonne. Coal prices, understandably, have been projected to grow at 0.5 % p.a. only in constant price terms and 2.6 % in nominal terms. How do these compare with current Thar coal prices estimates of USD 5.5 per MMBtu (40-50 USD per tonne). A food for thought for those who would like to keep Thar coal attractive and thus competitive.

Table 16 Energy Prices Projections (2015-2040) at nominal and constant prices of 2015 in USD

	2015	2025	2030	2035	2040	RoG (%)
Crude Oil Brent USD per bbl						
at constant prices of 2015	52	92	104	120	136	3.9
at nominal current prices	52	112	141	181	229	6.1
Natural Gas USD per MMBtu						
at constant prices of 2015	2.62	5.12	5.06	4.91	4.86	2.5
at nominal current prices	2.62	6.27	6.84	7.42	8.17	4.77
Coal at mine USD/MMBtu						
at constant prices of 2015	1.69	1.71	1.71	1.86	1.91	0.5
at nominal current prices	1.69	2.09	2.31	2.81	3.21	2.6

Source: US EIA Annual Energy Outlook 2016

Figure 6 Break Even Prices of Oil Companies

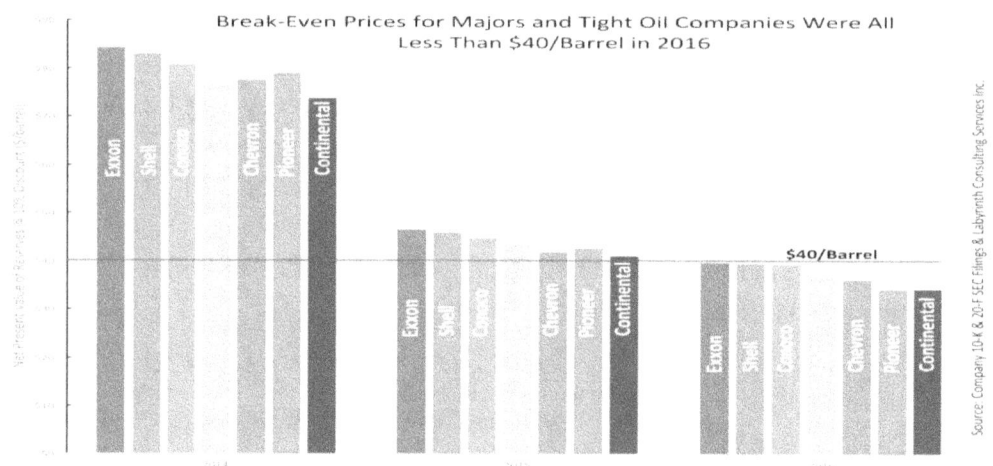

Source: Berman, Arthur (2017). Why Breakeven Prices are Plunging Across the Oil Industry < http://oilprice.com/Energy/Oil-Prices/Why-Breakeven-Prices-Are-Plunging-Across-The-Oil-Industry.html

8. Priorities in CPEC

Joint Coordination Committee (JCC)[5] completed its deliberation in China, in December 2016, with Pakistan side being headed by the Minister of Planning, Prof. Ahsan Iqbal. New projects were approved, mostly on roads, circular railways and industrial estates. This time the power projects were less in number. Every province, with the help of China, will establish one industrial estate, and one rail-based mass transit scheme. There was also a long list of miscellaneous provincial schemes, which many found slightly out-of-place in a larger framework such as of CPEC. It was a good omen, however, that Chinese have not been discouraged by the experience hitherto and have come with more. In this chapter, we draw attention to issues that merit some consideration for prioritization in the CPEC framework.

Heeding to the provinces' grievances, this time there were more consultations with the provinces. In fact the provinces themselves proposed their own schemes. Normally, such approvals are in-principles to be finally decided upon the outcomes of the feasibility studies. Financiers always give importance to the returns of the project, although recipients are not usually very sensitive on this as for them it is "Qarz ki Mai" (wine on credit as per ghazal of Mirza Ghalib). Provinces should not insist on unviable projects and should give importance to the financial viability. After all, it is they who would have to pay, unlike earlier CPEC projects of federal content and nature.

Every action has both positive and negative consequences. A matter of concern is the projected debt servicing liabilities and even imported fuel bill. One 1,000 MW power plant would generate foreign currency payments of 600 million USD per year. For 10,000 MW, it would be 6 Billion USD per year. In fact in the initial years, it may be 30 % more due debt scheduling issues. This would certainly create a lot of strain on Pakistan's balance-of payment capabilities, unless exports are raised concomitantly. Normal export increase prospects do not appear to be very high.

---

[5] The Joint Coordination Committee, chaired by Minister Planning, Development and Reforms, have representatives of all stakeholders including the four provinces and Gilgit-Baltistan

Developing Local Exports

It may be a good idea to create export projects meant for export to China. One of such possibilities is export of minerals, particularly, copper. Readers may recall the Reko Diq mining project, which became victim of the immaturity of a motley crowd of communists, nationalists, ill-informed armchair economists and above all an activist judiciary of the time. Some other enthusiasts raised unrealistic optimism for self-development and revival of the Reko Diq project, little realizing that it requires several billion US dollars of investment, technology, management skills and marketing network. Otherwise artisan mining with an axe and spade is being done for centuries. With a power Shovel now, everybody thinks he can mine anything. Local coal resources have been destroyed by poor practices of local miners.

A copper project of Reko Diq size can generate exports of more than one billion USD, thus supporting foreign exchange requirements of 2 coal power plants of 1000 MW each. However, legal situation does not appear to be clear. Some fishy negotiations have reportedly been done in the past with the original promoters who may have a claim of sorts. There are other large copper deposits in Chaghi outside the Reko Diq project area, which may be free of legal complications, where Chinese investment could be invited and should be invited soon. Wheeler-dealers may continue to revive Reko Diq in parallel. However, precious time should not be wasted waiting for revival of Rekodek. Baloch Nationalists could be educated on the benefits of the project to Balochistan. Resources lose value over time, as technologies change and mineral resources lose importance. Coal is getting out of fashion as there is a global shift towards green energy, and usage of other metallic minerals has gone down. Resource companies and countries hasten to exploit their reserves, as one sees in case of oil. The time for Copper exploitation is now. One may repent later.

Chinese interest in Pakistan (Balochistan) did vane away in the past due to law and order problem in the pre-CPEC era. Now with CPEC framework and Gwadar port activities, there is a new situation. China would be interested now, especially, in the context of repayment of their invested capital in other projects. In fact with the integration of transport infrastructure, even mineral sector of KPK has become attractive for China.

Prioritizing Local Content and Manufacturing

Local industry has suffered, under the fast track power projects, for understandable reasons. It takes usually more time to get components manufactured locally in Pakistan. However, with the completion of 10, 000 MW projects, emergency would be over and in the new projects, some delayed delivery by local manufacturers can be tolerated or absorbed in the overall schedule.

Conversations with Chinese have revealed that China is aware of the issue and that they want to assist in industrialization and development of local technological infrastructure in Pakistan. China has done it earlier too. Major technology ventures such as HMC –HFF, Railway Carriage and Locomotive factory and nuclear power etc. in Pakistan has been reindustrialized for a variety of reasons: the new trade order and a naïve enthusiasm for private sector investment in high-tech industries. The latter has not happened. Even, there have been no takers for projects on prioritization list. Those who took over, later closed down factories and sold the land. Pakistan Steel has been destroyed in the process of privatization. More or less,, the same has happened to HMC and HEC and others.

Countries like Iran, Turkey, Indonesia and Egypt have developed significant power plant equipment manufacturing capabilities. Power requirements usually double in a decade. Earlier, lack of demand could have been cited as a reason for not developing the plant equipment manufacturing capabilities in Pakistan. However, this isn't the case any longer as the economy is expected to grow at 7% by 2020. Thus it is imperative that such projects be initiated under or outside CPEC framework.

Developing infrastructure for Water Conservation and Efficiency Projects

Water conservation and efficiency projects should receive priority, as Pakistan is already a water stressed country. An evidence of the problem is already visible these days. If sufficient attention is not given, we may be facing a similar crisis as we have faced in case of electricity. Pakistan has practically one water storage dam of a significant capacity, namely, Tarbela the capacity of which is being continuously reduced due to sand settling. Bhasha dam has been inaugurated several times over in the last decade or more but it continues to suffer from international politics. The only possibility of its implementation appears to be through China. Time and politics keep changing. The window of opportunity is now. Time, resources and opportunity should not be squandered away. Present government has taken serious interest in Bhasha. 10 Million USD have been spent in the land acquisition which is usually a long and arduous process. Bhasha is ready to be implemented. Proposals for financing Bhasha under local financing have been forwarded, which appear to be rather too optimistic. We should put our priorities right and put Bhasha on a higher priority than it appears to be at present. It would be highly unfortunate that we delay Bhasha to a point, when Chinese announce that the budget is over.

9. The Conundrum of Single Bid

Over the last few years, we have seen many projects being implemented under single bid approach, which in some cases has resulted in rather excessive costs. However, if the funding is available under a bilateral arrangement, there is possibly no escape from single bid projects. Competition can be organized if funding is independent of supply, which is only possible when the funding is from third sources or multilateral institutions.

IFIs or multilateral institutions are not the best choice either as they do not have the kind of resources that have been made available under CPEC. There is a long processing cycle in IFIs and the funding is associated with conditionality and agenda that may be compatible with international frameworks, but often in terms of scope and timings, not realistic or compatible with the day to day running requirements of the country. Moreover, they always require some kind of surgery. Irrespective of these peculiarities, as we have mentioned, IFIs have to loan to many borrowers and thus cannot cater to the required level of funding.

The issue is not just limited to CPEC, it is common to all bilateral funding such as from Russia, UAE etc. in the past. EXIM Bank and the other US banks partially have attempted to tackle this issue, wherein they require agencies competition among the bidders, although restricted to the American suppliers. Perhaps Chinese can be prevailed to do the same. They have a large economy and many suppliers of single products. They are encouraging and enforcing competition in the domestic sector. It may not be a bad idea to try the same in external markets.

Comparative Analysis of Projects with Single Bids

I will give you two or three examples on which I have data and understanding of the projects. Gawadar-Nawabshah gas transmission pipeline is planned to be built under CPEC at a cost of 1.326 Billion USD. It is of 42-inch diameter and a length of 700 Km. Average cost per km, in this case, comes out to be 1.9 million USD per km. Another transmission pipeline is to be built (South-North) by Russians, connecting Karachi to Lahore with a length of 1,100 km. diameter 42 inches and a throughput capacity of 420 billion cft per year. Its capital expenditure has been estimated at 2.5 Billion USD, which comes out to be 2.29 USD per km (three times the rate in India. SNGPL has completed a

111 km project recently at a cost of Rs 9.187 Billion, yielding a unit CAPEX rate of 820 Million USD per km). Russians demanded, reportedly, 1.20 Billion USD but settled for 85 cents per MMBtu (which is still almost twice the required one). In India, the comparable tariff is 60 cents for a 30% larger pipeline; Gail India asked for 1 USD/km but was awarded 60 cents by the regulator PNGRB, which is still under consideration to be brought down significantly. By comparison, SNGPL operates a network of 7,756 km of transmission and distribution network of 94,263 km and delivers 434-600 Bcft per year of gas in just 100 cents per MMBtu, i.e. only 15 cents more than the proposed Russian pipeline tariff. Most of SNGPL annual expenditure goes into salaries and distribution. Transmission component alone may not be more than 20 cents and in actuality be even less. Admittedly, one cannot expect new project costs with the old ones, but there has to be a limit and reasonableness. A difference of 100 % or more even among new projects should be a cause of worry to all.

It will all add up to make RLNG very expensive; terminals, pipelines and many intermediaries with their extractive pricing. MPNR has traditionally been a generous agency. It has been in earlier periods awarding oil pipeline projects with a ROE of 25% (unprecedented anywhere).

One apparent reason is that rather excessive CAPEX is demanded by the single bidders i.e. almost twice or even more than the local cost. Typical regional rates are around or even under USD 1 million per km for comparable projects. Admittedly, there is normally a wide variation in CAPEX of pipelines as no two projects are the same. Terrain can lead to substantial differences in cost. Gawadar Nawabshah has an inhospitable terrain of about 100 km and the law and order problem adds up to the costs and risks. The only way to be sure of real prices is to have competition.

In India, a comparable project (Dhabol-Banglore, 1386 km, 36-inch diameter) has been built at a cost of 670 million USD in 2012-13.This gives a unit CAPEX of around 500 million USD per year. Accounting for diameter difference and inflation since 2012, one may add 50% to it. Perhaps more relevant project being currently implemented by GAIL India is Jagdish-Haldia pipeline (36-inch diameter, 950 psi, 2538 km) at an average cost of 0.717 million USD per km. It is a fairly complex project with 46 river crossings, 17 railways crossings and 14 state highways crossings. Tariff for long distance pipelines hovers around 60 cents in India and efforts are being made to bring it down significantly. Gail India applied for a tariff of one Indian Rupee but got only half of it from the regulator and all of it in local finances. In our case, nothing of the sort applies. It will all go abroad.

On HVDC project, i.e. Matiari –Lahore electricity transmission line (660 kV, 4,000 MW, 878 km), Chinese company kept demanding high CAPEX of more than 2 billion USD and a tariff of more than Rs. 1 per kWh (at a load factor of 95 % which may not be achievable and at the more usual rate of 60-70% , it would be correspondingly higher) as opposed to existing tariff of NTDC of 33 paisa per kWh. Through protracted negotiations, an agreement has been reportedly signed at a mutually agreed CAPEX of 1.5 Billion USD. One is not sure if an agreement has been reached on tariff as well which has been determined at around 77 paisas per kWh by NEPRA. Reportedly, this has caused considerable tension between NEPRA and the line ministry.

We have given examples for the energy sector, which should not mean that all is well in other sectors. The challenge at hand for the concerned government agencies is what should be done to avoid the curse of monopoly. We neither have the money nor the technology, and the job has to be done. But would we be able to make repayments of the loans and take-or-Pay payments in US Dollars? Will our consumers be able to pay? Would our industry remain competitive, if unaffordable charges and tariffs are imposed on them? This would have negative impact on our exports, which would further erode our financial capacity. The answer is that in some cases, exceptions may have to be made and bitter pill would have to be swallowed. This cannot be, however, made a rule. There must be some untied loans as well. For example, given the money, pipelines can be erected by our local companies much cheaper and rather quickly, as long negotiation cycles are not involved. Or some JV or consortium type approach can be adopted to bring down the cost. The idea is not to criticize the relevant bureaucracy or the ministry. The problem is structural and this is the curse of the single bid.

What can be done?

The issue merits serious considerations in a consultative mode at G-to-G level as it may not be sustainable in the long run. After all, repayments have to be made, projects have to remain financially and economically viable resulting in user-charges that are affordable by the consumer. The government is an intermediary only; the final payer is the consumer. We are seeing that in some cases, the resultant consumer prices and tariff appear to be quite on the higher side. While individual cases may be taken lightly on a case-to-case basis, the cumulative impact can be disastrous. You may have facilities, but consumers may not be able to use them or reduce their relevant budgets. A circuitous spiral into recession and stagnation can ensue resulting in default and cessation of loan giving cycle. In this way, the interest of the supplier countries would also suffer. It would be in the interest of both the lender and the borrower to think of ways to minimize the vices of unfair pricing by the commercial companies, which normally conduct business based on profit maximization on a transaction-to-transaction basis rather than having a longer term perspective. They don't generally believe in the long run, arguing that "in the long run we all die".

The issue is not simple. It is not always easy to determine the right price without competition. In fact, in quite a few cases, competition is employed only to discover price. Cost can be determined if expertise is available, which is usually not available in the borrower country. One may not need to borrow if one was developed enough. Third party advice is either expensive or is not timely. Tendering requirements consume a lot of time, militating against hiring of third-party advice. Vested interest also works against it and manages to pervert the process.

However, it is possible to minimize the losses or inadequacies by installing a rigorous examination and scrutiny system for single bid projects. Partly, the system has worked due to the presence of regulatory agencies, which attempt to act as neutral empires in determining costs and tariffs. But all sectors are not covered by NEPRA and OGRA. Also, regulatory agencies are obstructed or to put it lightly are disliked.

Strengthening Planning Commission

Planning Commission could fill in the gap and provide third party role but it has lost credibility over the years and the vested interest magnifies its capacity or capability problem for understandable reasons. Planning Commission can serve as a third-party institution as it is not the one, which originates proposals. Over the years, its role has been systematically or otherwise curtailed. In its current state, it may not be able to do this efficiently and adequately. Necessary organizational changes can be brought in as some proposals are reportedly on the cards or in the mind of the leadership of the Planning Commission. Planning Commission needs to have induction of several scores if not hundreds of qualified professionals with at least one-third of those being PhDs to be able to do a good job. It has to spend up to 1 % of the project cost in project evaluation hiring external and independent consultants as opposed to under-staffed and under-paid situation that it is in. Instead of opposing the Planning Commission or proposing its closure, steps should be taken to institutionally strengthen it and remove its inadequacies that are there. It can serve as a think tank and as a third party evaluator of proposals, not necessarily limited to CAPEX approval.

Indigenization

We have been arguing for indigenization in this space. Indigenization or co-production in the form of JVs or otherwise could reduce and control costs and develop references for contracts in other countries. Pipelines are not built every day and precious opportunities have been wasted in this respect. Lender's interest can be taken care of through buying inputs like pipes, electrodes and other consumables and electronic packages that form normally 65-70% of the real costs. Easier said than done, as lenders would not like to develop potential competitors' credentials. The easiest thing in the energy sector is laying pipelines. There is a considerable local know-how and expertise. A local pipeline company could be floated for these projects, which could handle these projects in a JV approach. After all, MPNR is floating all kinds of sundry companies like Pakistan LNG Terminals Ltd. But that would perhaps be a client company for lording over and not the kind of company that we are proposing here which requires hard work.

Allowing stronger regulatory role

To be fair, all is not that bad. RLNG based power plants are being built with very low CAPEX. However, there was stiff competition, and it was not a single bid. Coal power plants are being built under a reasonable tariff developed and scrutinized by NEPRA. Upfront tariff, independently arrived at by the regulator eliminates the need for competition to a very large extent. This was made possible by involving NEPRA and not avoiding it.

Unfortunately, MNPR does not like its regulator OGRA and prefers to have a decision made by ECC in haste. It has not yet taken the pipeline cases to the scrutiny of OGRA, which could have enabled it to get a better deal from the suppliers. Had OGRA got some teeth, it could have announced upfront tariff (indicative at least if not mandatory and final) to guide and discipline the process.

In competitive bids as has happened in case of LNG terminals, regulatory process may not be required. However, in case of single bids, it is highly objectionable to avoid regulatory process. It could have been done much in advance to save time. There ought to be a difference between running a company and a ministry. Even in case of public listed companies in the west, some compliance is required these days in procurement processes. Negotiating committees comprising of government officials are not a substitute for regulatory process. In fact, negotiations have to take place after a verdict has been received from the regulator. As per Planning Commission guidelines, a feasibility study is to be completed by a third party. The problem is that if the single bidder is supposed to conduct feasibility study and the cost estimates also are to be done by him, how can there be fair pricing and on what basis an objective price negotiation can take place?

We have provided comparative examples from India earlier in the chapter, which indicates how the regulatory process drastically cuts the demands and comes up with realistic tariff. Institutional controls and third-party involvement is an asset, not a liability. If the suppliers know that they have to convince only one agency, they build in more cream in their demands. So let the institutions prevail and whatever institutional arrangements are there may be allowed to work so that such conundrums are handled at least partially. Let us work together creatively and innovatively to eliminate the inadequacies that happen to be there. If investment has to come from China only, as it appears, some bilateral institutional arrangement (e.g. bilateral oversight, competition among Chinese companies) may have to be made to prevent possibilities of excessive profiteering by the suppliers.

Loan Terms vs Competitive Bidding

Under tied bilateral loans, there is no competitive bidding, except in US EXIM Bank lending. Usually CAPEX, and EPC prices are high by 20-25% or even more. However, loan terms are kept soft. In Table 17 we provide some data in this respect. We see that interest rates are quite low as opposed to CPEC energy projects, which effectively come out to be 5.5-6%(LIBOR+4.5%),in addition to hefty Credit Insurance(Sino-sure) of 7% and financial charges of 3%.Pehaps,Chinese had no fault in it. It is NEPRA policy. However, nobody would have stopped them, had they charged competitive rates. Also had NEPRA adjudicated wisely and fairly, there was no body stopping them. There is nothing on record that indicates some altercation between NEPRA and the Ministry of Water and Powers or other agencies such as EAD or Ministry of Finance. NEPRA interest rates are in the context of general private sector commercial loans where there are no G-to-G tied loan arrangements. In the instant case, there is double or triple jeopardy: high CAPEX, no bidding, no concessional lending, imposition of extras like Sinosure and heavy financial fee. As it is NEPRA's risk margin of 4.5% over LIBOR is a bit high. It used to be 3 % several years ago. Admittedly, one is wider in the hindsight and it was done in haste. But now, there is enough time and the loan terms should be negotiated fairly and separate upfront coal power tariff rates be issued for CPEC projects.

Table 17 Bilateral Project Loan Terms in South Asia

	Year	Project	CAPEX(MnUSD)	Interest Rate
China Sri Lanka	2007-12	Coal PP,300 MW	450	3.3% composite
China Bangladesh(Garoshal)	recent	NGCC,450 MW	421.25	LIBOR+3.5%
China-Bangladesh	recent	Coal PP275 MW	330.52	3.99%
India-Bangladesh(Rampa)	recent	Coal,1320 MW	2000	LIBOR+1.75%
China-Bangladesh	recent	General	20000	2%
CPEC-Gawadar Pipeline	2016	pipeline	2000	2%
CPEC-Electricity Project	2015-17	Coal PP and other		LIBOR+4.5%
CPEC-Roads and Transport	2015+	Road Transport	8000	2%
Source: Various international news				

Source: Compiled by the Author, IEEFA, Planning Commission

PART II: RENEWABLE ENERGY, CONSERVATION AND CLIMATE CHANGE

10. Are We Consuming Energy Efficiently?

Due to depleting fossil fuel resources in the world and the climatic effects of burning these resources, there is a lot of stress these days, on countries, to reduce energy consumption. In developing countries like ours, who have only begun on the road of industrialization, the only option is to be in harmony with the world community is to improve our energy efficiency by controlling wastes and producing and consuming less CO_2 producing electricity. Thus energy efficiency is to be better understood by all of us to be able to better contribute towards achievement of this objective.

In earlier days of development and industrialization in the world, energy consumption was considered an index of development. It still is in many ways. United States, China, Russia are the largest energy consumers. However a country like Bangladesh may be consuming more energy than Denmark, but it is the population in this case which is driving the difference. Per capita consumption of Denmark, on the other hand, is certainly several times higher than Bangladesh.

On the average today, to produce 1 USD of GDP, the world is consuming 0.14 kg of oil, the latter costing about 5 cents. Simply put, to produce, 1 USD of GDP, one has to buy or consume energy worth 5 cents. We may also, say, that world average intensity of GDP is 5%. This may not however, be confused with energy efficiencies of power plants and other devices where the efficiency may vary from 30% to 90%.

Energy consumption in a country may be higher due to population, structure of economy (producing raw materials, extractive industries etc.) and energy inefficiency. World Bank has compiled Energy Intensity figures of most countries of the world. They have tried to measure as to how much USD of GDP is produced by consuming 1 kg of oil by taking GDP of the country at PPP of 2011 and dividing it by the total energy consumed by that country or economy.

In terms of this index, China and Canada produce least of GDP (only 5.40 -5.90 USD) out of 1 kg of energy consumption. Both China and Canada have large raw materials and extractive industries, which are highly energy intensive resulting in lower GDP-Energy

Index. European countries like Germany, Spain, etc. have twice or even higher GDP/Energy index (10-13 USD), because of their high value added economy with small raw material component. Among the European countries, Italy has the highest GDP-Energy Index, due to relatively expensive energy in that country, dictating energy efficiency and discouraging energy industries. U.S. and Australia lie in between the two afore-mentioned group of countries, having the characteristics of both the groups.

One may be surprised by a high GDP –Energy Index (13.20) of Bangladesh. Is it the most energy efficient country? Not at all. This represents the situation of a subsistence economy having least energy consumption. Mauritius and Mayanmar have comparable figures for the same reason. South Sudan has perhaps the highest GDP-Energy Index of 30 representing perhaps the least of development. Singapore at 16.10 certainly does not indicate under-development. It is perhaps the most energy efficient country among developing country. However, its trade intensive economy, moderate climate and smaller area has enabled it to have higher GDP-Energy Index than Europe, USA, China and Australia. However, there is one common feature in almost all countries, which is of increasing high GDP_ energy Index over the years, as the average value has gone up from 5.5 in 1993 to 7.7 in 2013 as shown in the figure.

Figure 7 Energy Intensity (GDP per kg of Oil) of Selected Countries

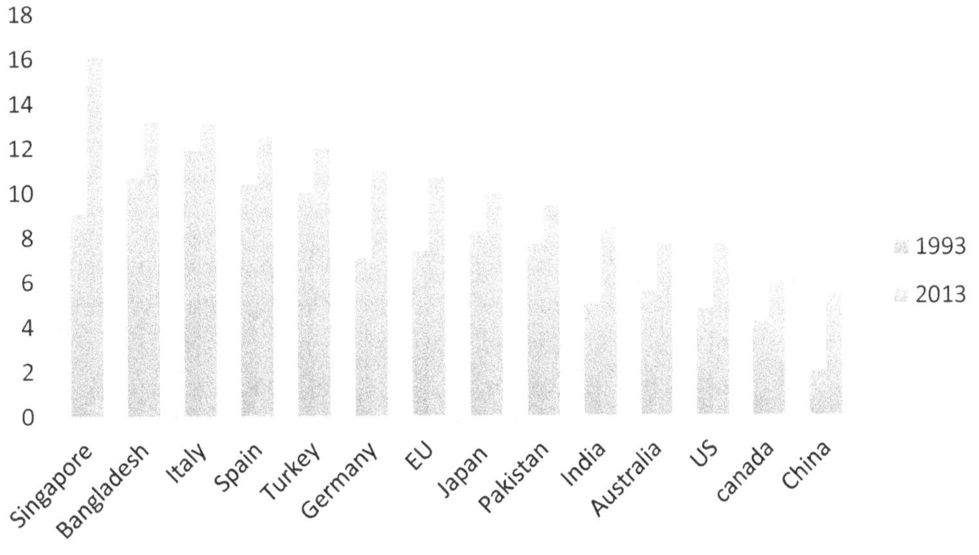

Source: World Bank

With this background, let us see where we stand on this issue. Pakistan's GDP-Energy Index has increased from 7.7 in 1990 to 9.4 in 2013 indicating an improvement of 22% in a period of 22 years. In the same period, India has shown an improvement of 68%, from 5.0 to 8.4; a rate of improvement of three times (3% p.a.) than that of Pakistan. All of this may not be due to energy efficiency measures, but may represent contribution of more value added industry than the extractive low valued added industry earlier.

About half of Pakistan's CO_2 emissions originate from the Energy sector. Electricity demand in the last few years has seen an exponential growth with a major portion of this growth coming from the non-productive domestic sector's increased electricity consumption. On the other hand, the proportion of electricity consumed by productive sectors, including industrial and agriculture, has been steadily declining because of their lower rate of growth. The share of domestic electricity consumption went up from 10% in 1970 to 43% in 2010 whereas the industrial share, during the same period, decreased from 46% to 24%[6]. This growth of the domestic sector has largely been fueled by the high subsidies the Government has afforded to the residential consumers undermining the financial viability of the power sector. If done nothing, the domestic sector will continue to grow exacerbating the demand supply mismatch and further straining the growth of the productive sectors.

The present Government, to its credit, has taken various measures to tackle the prevailing energy crisis, however primarily focusing on increasing the energy supply. On the other hand, not much work has been done on reducing the demand for energy. Improving energy efficiency will proportionally improve our carbon rating. There is a lot of scope in increasing energy efficiency. According to estimates given in various studies and surveys, lighting, space cooling and heating are two major electricity-consuming categories amongst households in Pakistan. These two basic uses account for two thirds of total household electricity consumption. The Government, few years ago, initiated a campaign to provide energy savers to households, which significantly reduced the use of inefficient tube lights and incandescent bulbs consuming at least four times energy as Compact Fluorescent Lamps (CFLs). On the average, a 25 Watt lamp (CFL) has replaced a 40-60 Watt conventional lamps, improving energy efficiency in the lighting sector by more than 100 %. Now LED lamps have arrived which would further improve efficiency. Replacement of CFL by LED would also eliminate the mercury problem caused by CFL. LED penetration at this stage is comparatively low. There have been proposals to remove all duties and taxes on LED imports to make the LED prices more attractive and

---

[6] Power System Statistics 2013-14

affordable. Government agencies, provincial and federal should launch a major program towards LED installations in their buildings and facilities. Fan industry has also improved its efficiency ratings. Leading manufacturers have been producing fans of fewer than 80 Watts.

It is the mandate of National Energy Efficiency and Conservation Authority (NEECA) to enforce Minimum Energy Performance Standards and labeling program covering appliances and various types of machinery. NEECA, in collaboration with Punjab Energy Efficiency Conservation Authority (PEECA) is in the process of setting up testing labs for fans, air conditioners and motors at various academic institutions for issuing energy labels for these products. Based on the tested energy efficiency rating of the product, their manufacturers will be granted permission to affix "Pakistan Energy Label" on their product. A higher star rating indicates better energy efficiency rating i.e. for the same amount of energy used, the product will deliver a better output.

Figure 8 Pakistan Energy Label

Source: NEECA Website

In industries, financial and energy savings are attainable through improved operation of tools and machineries and through efficient production processes. These measures mostly require behavior change and can yield substantial savings by undertaking very low cost investment. The savings can be multiplied by replacing old and inefficient machinery.

Buildings are also significant consumers of energy though their performance can be improved by strict enforcement of building energy codes. Pakistan Energy Council launched the building energy code in 2014. The code sets thresholds for building energy consumption energy consumption through design and construction standards that apply to energy systems, equipment, and the building envelope. After the enactment of Energy Efficiency and Conservation Act, NEECA and provincial energy departments have been entrusted with the task to ensure that these building codes are strictly complied with.

Planning Commission with the collaboration of UNDP have initiated SE4All program with the aim of doubling the energy Efficiency by 2030.A National Action Plan is being developed which is expected to bring all relevant organizations and institutions under a common framework. Indeed a lot can be done and is being done, although its pace can be improved under a fast track program. International community including (World Bank and Asian Development Bank) is quite enthusiastic in providing soft funding for projects and programs in this sector.

11. Towards a Realistic Climate Change Policy

Climate Change is a catastrophe that can be avoided by combined human action and cooperation, as human civilization has been able to tackle many issues that threatened its well-being and survival in the past. However, what level of effort and sacrifice has to be made by individual countries has remained a matter of contention.

Climate Change is caused mainly by the accumulation of Green House Gases (GHG) such as CO_2, SO_x, NO_x, and Methane etc. Predominantly, it is CO_2, which is a product of combustion of fuels and biodegradation of organic materials. These emissions lead to rise in temperature of earth and atmosphere affecting metrological balance and melting of glaciers etc. causing droughts, floods, hot summers, and rise in the ocean level threatening inundation of many coastal cities like Karachi and others.

Figure 9 Emissions Intensity (per GDP-PPP)

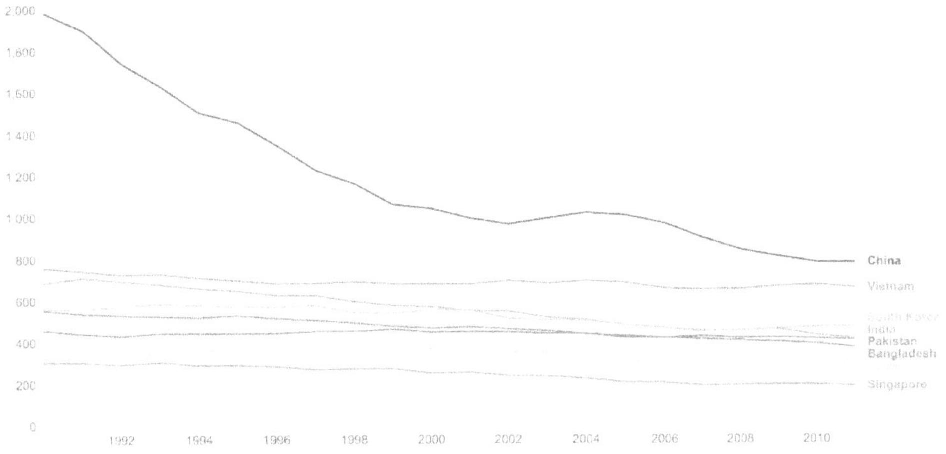

Source: World Resource Institute

Pakistan's contribution to the problem (emission levels) is miniscule indicated by its rank of 135 commensurate with its ranking in many other development indicators. The following figure plots the emission intensity per GDP (ratio of greenhouse emissions to GDP) of several Asian countries. As seen, Pakistan's emission intensity is low compared to countries like China, Vietnam and South Korea.

Energy consumption and thus emissions have long been considered as indicator of development and thus any limitation on it appears to be a limitation on development itself. For developing countries including Pakistan who are at best at an early stage of development, the demand of limiting emissions is akin to resisting and opposing their development .For developed countries, perhaps development peak has already occurred and thus action and sacrifice can be made on their part. Being technologically developed, it is easy for them to rearrange their resource inputs and reduce emissions which activity can in fact boost their economies. Developing countries are not able to do this by themselves without technical and financial assistance from the developed world. Even if such assistance comes through, most of it may be loans and very little grants leading to indebtedness.

Table 18 Comparative INDC Targets of Various Countries

Country	Target (% reduction)	Reference Year	Target Year
Iran	4	BAU 2030	2030
Egypt	None	-	-
Argentine	15	BAU 2030	2030
India	35	GDP basis	2030
Thailand	20	BAU 2030	2030
Turkey	20	BAU 2030	2030
Chile	30	GDP basis	No target year was given
Bangladesh	5	BAU 2030	2030
Indonesia	29	BAU 2030	2030
Pakistan	20	BAU 2030	2030

Source: UNFCC

International discussions and even polemics have occurred on whether the load has to be borne by developed countries alone or some effort may also come from the developing countries .The conundrum is that although individual countries emissions are miniscule in per capita terms, combined and in absolute terms the developing countries contribution as a whole to the problem aggregates to a significant number. It has therefore been agreed that they will estimate their own share and level of effort and declare the ensuing targeted emissions as their plan of action and submit such affirmations in the form of Intended Nationally Determined Contributions (INDCs). Table 18 provides INDC targets of a number of countries.

Developing countries including Pakistan have been in confusion as to what should be their target and what kind of commitments can be made and even estimating the emission levels and possible reduction without damaging their economies has been beyond most developing countries. India submitted its INDC last year and Pakistan could do it only last month. The issue still remains as to whether any reduction in emissions can be targeted, as our current emissions or even projected ones say of 2035 are going to be too low. Reduction in emissions say from 2035 level may tantamount to limiting our development. Pakistan's Climate Change policy makers, perhaps under pressure of announcing some commitment and to be the part of international process and to be able to benefit from Mitigation assistance packages that are to ensue , have submitted INDCs , committing to a reduction of up to 20% of its 2030 projected GHG emissions subject to availability of funds.

In my view more creativity and effort should have gone into estimating and decision on INDC commitments. We could have reviewed INDC of other countries, especially, of the region, particularly India. India has not agreed to any absolute limitation on its emissions at all. India has only agreed to reducing its emissions per unit of GDP by 30% by the year 2030.This is reasonable as this amounts to improving energy and emission efficiency by implementing conservation policies. Pakistan should also make its INDC submissions on these lines. Things are not at a level of finality at this stage. There is still time to come up with emission reduction targets and approach that is consistent with our unrestricted rights to growth. We can show our sincere commitment to the international community by agreeing to promoting conservation, renewable energy induction and development of emission sinks like forestation. This would be taken more seriously than a kind of commitment that eventually may not be implementable.

Unfortunately, Ministry of Climate Change (MCC) is in its infancy, although it is trying to do a lot. A Climate Change Law is in offing, and a Climate Change Authority is being made. However, it need not reinvent the wheel. Climate Change is a multi-sectoral issue .In this context Planning Commission can come to the rescue of MCC. Planning Commission has multi sectoral resources, which can be marshaled to support MCC. For example, they have a full-fledged Energy Wing, which is building an integrated Energy Model. Energy contributes to 50% of GHG emissions in Pakistan. Emission module could be added to the Energy model without much difficulty. There are many other ways and means through which PC can be of help. For Planning Commission, yet it is another opportunity to meaningfully contribute and improve its image in a target area that is rapidly evolving. In fact it is through such opportunities like Climate Action and Inter-provincial integration that Planning Commission can renew itself, legitimizing its very existence, which has emerged from many quarters even from within government departments.

Concluding, although our contribution to the causes of Climate Change is insignificant, the consequence and catastrophe that is threatening us is very dreadful. We have been declared third most endangered country in terms of Climate Risk, the impact of which has already beginning to tell on us in the form of floods and hot summers. There is threat of droughts and floods alternating. And our largest city Karachi is on the coast facing risks from inundation from ocean and many others. We have to plan and prepare for mitigation steps and infrastructure in order to be able to deal with the challenge adequately. Crisis is an event not planned for.

12. Competition in Renewable Energy

Utilities have been considered as natural monopolies as these used to be integrated companies combining generation/supply, transmission and distribution. These days integrated utilities are rare (K-Electric should have been dismembered before allowing sale of its majority shares to a foreign company, as it would make rationalization and restructuring difficult) and generation of electricity and supply of gas is universally competitive. Distribution companies buy electricity at competitive rates largely determined in Energy/ Commodity exchanges in a stock market like situation. In developing countries, however, there are regulatory agencies like NEPRA/OGRA who determine the prices-fair prices both to producers and consumers. Many people in the country are skeptic of competition as has been indicated by frequent raids of even a docile Competition Commission of Pakistan. This has created justification for continuation of regulation.

Time has come to introduce some competition as we will see in the following the failure of regulation. Despite one's reservation against competition, which can be manipulated and hijacked, the world experience is that competition drives down prices more than regulation. Prices are discovered through bidding. Otherwise, there is no more accurate way of knowing the prices of inputs and outputs. Cost estimates by regulators or bureaucrats may be faulty, in favour of supplier or the buyer. Interestingly, in our case it appears, that mistake or faulty assessment has never been done against the suppliers. Finally, in regulation, everybody is equal: all companies irrespective of their balance sheet, experience and credibility, get a risk spread of, say, existing rate of 4.5 %. Everybody manages to get a supplier and a lending agency with this kind of liberal spread. In a competitive situation, better companies are able to get better borrowing terms and thus are able to offer better prices and get the contract. In regulated environment, all and sundry form a queue vis-à-vis the buyer and the latter has no clue as to whom to sign PPAs with and thus the premium.

It augurs well that NEPRA, after observing all that has happened in the past and facing difficulties in arriving at the right tariff, has come out in support of competition, although a bit late as they say, *dair ayad, durust ayad*. NEPRA has issued guidelines for competitive tariff for solar and wind projects. Reportedly, AEDB has responded by asking more time as auctioning is new. However, consultants can be invited to assist in reverse auction. Real reason for procrastination appears to be different. Firstly, CPPA, the power purchaser is already full of 1,000 MW PPAs it has to pay for. Secondly, CPPA is already under financial load of paying for the expensive solar power capacity of 400 MW or so. Thirdly, there are transmission issues possibly, as there are problems in transmitting renewable energy, which is often non-steady if not unstable.

There was and is no need that solar and wind tariff be kept so high as compared to other countries under vendor influence, flawed regulatory practices and reliance on public hearings without making use of good third-party expert advice. Although regulation is almost always less efficient than competition, regulation could have been better conducted. The above should not be construed as indictment of the present leadership either in government or the regulatory agency in question. It has been a common problem over the years. Let us look forward to better days and an appropriate tariff under competitive conditions.

Falling Renewable Energy Prices

Remarkable events have been occurring since last one year or so in the area of solar energy pricing. Projects have been announced, one after the other, with unbelievable prices like 3 cents per kWh for a project in UAE and many other projects elsewhere announcing an average of 5 cents solar tariff.

In February 2017, solar power prices have come down as low as 5.30 US cents per kWh in India. Wind Power could not remain aloof, and its bids have come down to lower prices of 5.00 US cents. Recent competition in Solar PV has taken place for supply of various project capacities.

In 2010, solar energy was contracted at a global average price of almost USD 250/MWh, compared with the average price of USD 50/MWh in 2016. Wind prices have also fallen, albeit at a slower pace, as the technology was more mature in 2010. This is the magic of auction or competition. Wherever auction/competition has replaced regulation, there has been drastic reduction in energy prices. Brazil, Mexico, South Africa and others have experienced the same. There are extremes of UAE where Solar PV auction rates have been as low as 3 cents. Long term PPAs in the U.S. have been signed at 2-2.5 cents. However, there is a subsidy in the U.S. in the form of PTC (Production Tax Credit), which these days amounts to 1.50-2.3 cents. Thus adding this, real price in the U.S. for Wind Power comes out to be 3.5-4.8 cents.

Figure 10 Average Prices Resulting from Auctions, 2010-16

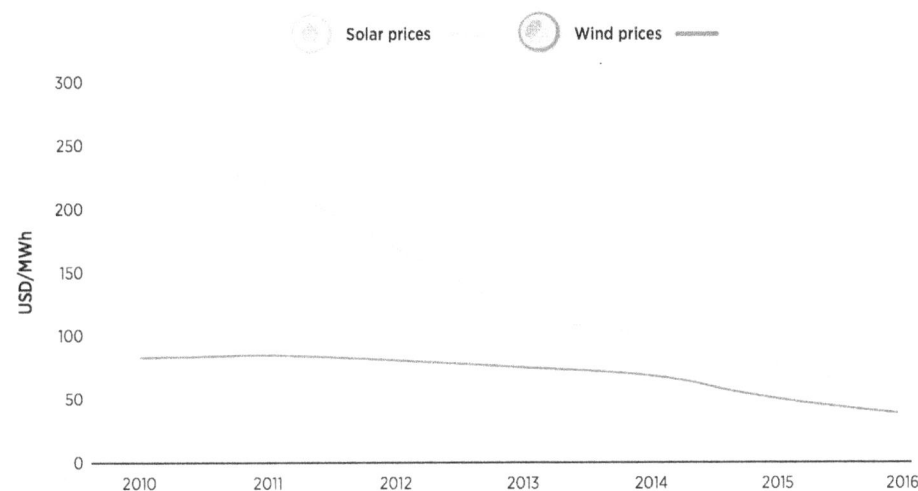

Source: IRENA 2017, Renewable Energy Auctions

Competitive Bidding for Renewable Energy Projects in Pakistan

NEPRA in January this year announced its determination on benchmark wind Power tariff as 6.75 Rs per kWh (6.41 cents) for foreign funded projects and 7.73 Rs (7.34 cents) for locally funded projects. NEPRA had invited comments in June 2017 providing the following as reference tariff; Rs8.61 for 100 % local debt and Rs 8.20 for 100 % foreign debt. Local interest rates are higher than foreign, possibly even after adding depreciation. As a reference only, going rate for coal power in India is IRs 3.0 per kWh or 4.5 cents. As against 5.19 cents for new solar PV electricity prices in India, reportedly serious offers have been received in Pakistan of 6 cents. This is remarkable, as normally there is a significant difference between India and Pakistan prices.

NEPRA, in the past, has been awarding rather excessive tariff of Rs. 14-17 per kWh for wind power, while market rates elsewhere were around 8-10 cents, 60-75% higher. In 2012, the tariff was highest in the world at 15 cents, while in countries like Brazil, Turkey and India, it was between 7 and 8 cents. It should be noted that in India capacity factor is lower than in Pakistan; wind cost is inversely proportional to price and higher the capacity factor lower the price. The capacity factor in Turkey and Brazil is comparable for wind. Some reasonableness was brought as late as in June 2015; NEPRA reduced Wind Tariff to Rs 13.20 to Rs.10.65. Its recent determination is a considerable improvement over the past (only 40% higher than minimum in India). Similar case was observed for solar power projects. In 2014, Solar PV tariff in Pakistan was 17 cents levellised, which came down to 11.35 cents levellised by Dec 2015. On March 3rd, 2017, NEPRA issued its order/determination on Solar PV without any tariff, benchmark or otherwise. NEPRA has finally announced that, henceforth, Solar PV energy price would be determined through competitive bidding.

Why are energy prices high in Pakistan? Firstly, regulated prices are always higher than competitive prices. The reasons are, briefly speaking, capital padding often permitted in the country as an incentive, and padding in other costs as well. For example, O&M costs are almost twice, and insurance costs the same. NEPRA has been holding public hearings wherein mostly vendor interests are represented. NEPRA generally does not have recourse to third-party experts. It is only after so many years that it has realized this issue and has started requiring competitive tariff.

I have been arguing for correcting this situation, over the years, in my writings and public hearings held on the subject at NEPRA. The same has been elaborated in detail in my book (Issues in Energy Policy, 2014). Low and realistic energy prices and tariff are good for everybody.. Higher prices contract market, demand and sales, while low prices expand sales and market. Pakistan's competitiveness has been going down as compared to our region and elsewhere affecting our exports. I have discussed in Chapter 5 the vices and difficulties created by single bid projects. Competition can also be a charade, if not organized properly. The buyer has to know about the prices, market and the product. Possibilities of collusive bidding and unrealistic risk perception of suppliers and other factors can defeat the very purpose of competition.

The falling prices syndrome of wind and solar energy has not occurred suddenly and was not unpredictable. You can read my writings in various newspapers that I did between 2009 and 2014. Was it advisable to have installed QA solar and other power plants at a tariff of 17.85 cents per kWh in haste, almost three times the present prices? The fault lies with regulatory practices as well as with developers and the purchasers. NEPRA tariff for solar was higher by 75-100 %; the residual difference is due to general price reduction. One can be wiser in the hindsight. Indians have been buying much more expensive energy earlier than they are buying now. However, the difference in their case is much lesser.

Tariff for Wind Projects

NEPRA has announced a Tariff of Rs.6.77 per kWh. It is an upper limit for competitive bidding that is to take place. GoS applied to NEPRA to award this Tariff as an upfront tariff which has been understandably rejected by NEPRA. In the first place, GoS was not supposed to be led by vendor lobby into this at the expense of consumers. Provincial governments are acting as free-riders on the pooled pricing system wherein the deficit is paid by the Federal government as subsidy. However, it is in their own long term interest not to take it too far, as if and when, the sector is provincialized, it is the latter which will be bearing the brunt. As things are going, it does not appear to be very far when this happens. With increase in electric supplies, it would become impossible for federal government to bear the burden of subsidies and circular debt. Already, the stress is visible; there is difficulty in paying the IPP dues and IPPs are protesting, threatening to call sovereign guarantees. It would have been more appropriate if a discount (say of 20-25%) was offered negotiated by GoS on the maximum tariff determined by NEPRA for Competitive bidding purposes. It would have become more palatable and comparable with prices prevailing in other countries.

GoS request could not have been granted, as the upper limit was based on maximum possible costs. Tariffs are not awarded on maximum possible costs, but on typical or median costs. As an example, let us take the CAPEX assumed in that tariff of Rs.6.77.It has been take at 1.60 Million USD/MW, as opposed to the median market value of 1.2 Million USD/MW, which is more than 50% higher. This may be the price of European suppliers for European conditions and thus has been taken as a maxima. In Pakistan, Wind Turbines are coming from China, where the WTs are being sold at a price of 4000 RMB/kW, which comes out to be USD 600 per kW. If 100 % margin is added to cater for other costs, the installed cost does not come out to be more than 1.200 Million USD per MW. As an evidence, we provide in the adjoining, a price graph released by GoldWind, the oldest and the most credible supplier of WT in China. There are other sources and studies as well which indicate a similar price structure and this is also consistent with the results of competitive bidding in other countries as we shall see in the following.

---

1)Zorlu has filed a petition with NEPRA and has offered a tariff of 6 USc based on a capex of 1 million USD per MW and a ROE of 12 %.In India where investments are being successfully attracted at 12% roe. India's regulated tariff provided ROE of 16% but nobody pays attention to regulated tariff in India on solar and wind due to the competitive situation. Other interesting features of Zorlu's tariff application are only 1.5% of capex as non-epc cost and 0.25% insurance. Nepra has been awarding insurance rates several times higher than that. And Pakistani investors have typically demanded 25% of capex as non-epc or owners cost and Nepra has been awarding all that happily. But no more, Nepra wants competitive bidding. I and many others have been demanding Nepra reforms for a long time. ROE of 20% was awarded to Thar and in return KPK demanded (possibly through a NEPRA member representing KPK) similar returns and as did others. On the other hand, our industrialists complain high energy cost to be behind falling exports and all other consumers complain as well. Time to have a balance?

Figure 11 Wind Turbine Bidding Prices Stable

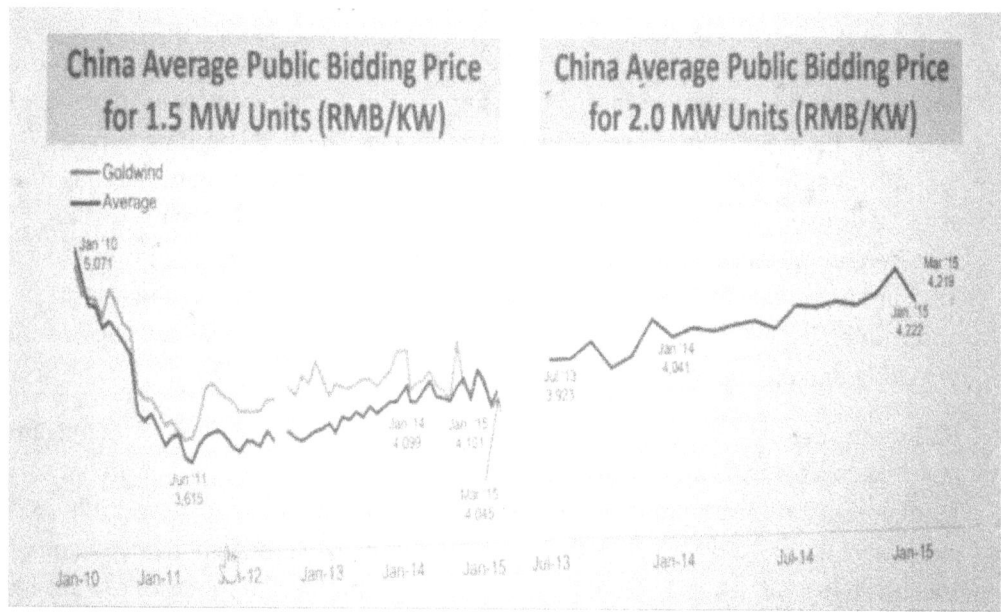

Source: Goldwind

There are several aspects to the problems and issues of the Renewable Energy. Firstly, investors have made investments in preparing projects and would like to implement their projects. Scores of projects are ready. This is like pregnant women ready to deliver, but not being allowed to do so. However, The REN industry of Pakistan itself is, at least, partly responsible for it. They have been asking high tariffs all along and have surrendered to lower tariff argument very late in the day, when it became a general knowledge among public that REN prices have come down so low. In Pakistan, for example, Wind power tariff was Rs 15.00 as opposed to around 7.5 Rs in Europe and India. If one sees the capacity factor vs. Tariff of Wind Power, one is startled. Highest tariff and almost highest capacity factor, when the relationship is in opposite direction. In India, the wind power tariff in recent auction has come out to be 5 cents, as against NEPRA's 6.77 cents. In the adjoining Table, we provide data on India's regulated tariff.

One would note that the highest capacity factor in India is 32%, while in Pakistan, the same varies between 35 t0 37%.If India tariff is extrapolated to Pakistan CF of 37%, one would get a figure of 3.84 Usc/kWh.

Table 19 Wind Power Levellised Tariff in 5 Wind Zones in India (FY2016-17)

	C.F.(%)	INR(Usc)
Wind Zone 1	20	5.9(9.44)
Wind Zone-2	22	5.37(8.592)
Wind Zone-3	25	4.72(7.552)
Wind Zone-4	30	3.94(6.304)
Wind Zone-5	32	3.69(5.904)
Pakistan-Jhimpir(extrapolated)	37	2.40(3.84)

Source: GWEC India Report(2016)

In other countries, developing and developed, REN auctions have yielded a Wind Energy price of 4.0 US cents, one cent lower than India. This is probably due to two factors; local content requirements and high duties on miscellaneous inputs; and secondly, because, India does not allow inflation indexation as opposed to other countries, meaning thereby that it is a fixed price. In the following, we provide excerpts from IRENA report. In the adjoining Table 20, we provide a comparison of Wind Power tariff parameters of two regulators, GERC India vs NEPRA Pakistan, which would let understand the difference in the regulated tariff system of the two jurisdictions. It is not clear what impact would be there of the taxation issues. GST has been levied recently on REN in India. In India tariff is denominated in local currency, in Pakistan it is convertible in foreign currency and the variations in foreign exchange rates are payable by the buyer. CAPEX in Pakistan is 60% higher and OPEX 100%.Similarly insurance rates in Pakistan exceed 1%, while in India it is 0.25 Rs per kWh. Some of the differences may have a rationale, while others may not be.

Recommendations:

1. Solar and Wind Power have tremendously improved their competitiveness; their cost of generation have almost become 50% of the fossil energy based power. In Pakistan, their induction has been partly obstructed by unreasonable demands of the investor lobby. Hopefully, competitive bidding will break that circular situation and the Solar and Wind power would be available at its true cost and prices. A fresh thinking is to be given for a large scale induction of these resources to be able to bring down the cost of generation.

Table 20 Wind Power Tariff Parameters: GERC India vs NEPRA

	GERC India	NEPRA-Pakistan
CAPEX	0.98 Mn USD/MW	1.6
O^M Cost	0.5 Usc per kWh	1
Escalation in O^M	5.72% p.a.	5%
Capacity Factor	24.50%	35
Project Life	25 yrs.	25
Debt Equity ratio	70/30	75/25
Term of loan	10 yrs.	12
Interest on Loan	12.3%(local)	5%(FE)
Risk Margin on Basic Rate	3%	4% on FE
Depreciation	6% for first 10 yrs	None
	2% afterwards	None
Minimum Alternate Tax	21.34%	None
Corporate Tax	34.61%	None
RoE IRR	14%	16%
Discount Rate	10.29%	10%
Tariff Levellised(Usc/kWh)	6.704	6.77

Source: GERC India and NEPRA

All existing power plans made by external agencies are outdated now due to the remarkable reduction in Solar and Wind prices under competitive bidding regime in many parts of the world including in our region. Least-Cost Generation Planning would certainly entail more induction of Solar and Wind Power than has been done in earlier workings. New studies must be commissioned. Intuitively, one could support an induction of 10000 MW in the period 2020-2030.This would come with the replacement of some coal and Hydro plants that are in current plans.

Solar Plan

2. Most energy and electricity planning has focussed on demand, supply and economics. Logistics and spatial planning has been ignored. Distributed generation requirements of Solar and water requirements of fossil power plants should be factored in a Spatial Plan which allocates power generation capacities and water withdrawal quotas. Solar Power should not be generated in the manner and style of fossil power. It should be generated

in a distributed manner.QA Solar Park may have been a good beginning. However, distributed solar generation close to population clusters should be preferred for which 30-50 locations should be identified in the proposed spatial plan. There is a strong case for planning a package of 5000 MW for 50 sites. Such studies should involve GIS technologies .Admittedly, this plan cannot be prepared in isolation. It should be associated with load studies. As a result, long hauls in Electricity transmission should be and would be discouraged saving energy and financial resources. Reportedly, HVDC project has been shelved due to good reasons, although, for this some coal projects had been introduced in Punjab in order to be able to obviate long haul transmission. Interestingly, coal transmission costs are almost the same as electricity transmission under HVDC.

Hybrid Solar-Wind and Others

3.A major shortcoming of all renewable energy sources is inconstancy. Solar available only during days, wind and hydro only in summer. And even during availability, there are variations. Its implication is that transmission and distribution capacities and investments remain underutilised for the period of unavailability. Also additional CAPEX is required for installing generation capacities for the unavailability period. Solar and Wind power plants can be operated in hybrid mode, utilising the space left between the wind turbines. Studies have shown that the shadowing effect of wind turbines on solar panels is limited to only 2%.India has already announced a Hybrid solar policy .However, the opportunities for Hybrid Solar and Wind appear to be only in Sindh, as wind corridor is available in Sindh and may be some parts in Balochistan. The latter can be a major beneficiary, as its most areas are beyond transmission grids. In KPK, there can be some opportunities for Hybridising Solar and small hydros. Sindh/Balochistan government should commission a feasibility study of a project and based on it a Solar Hybrid Policy should be developed.

13. Bhasha Dam: beg, borrow or steal?

There is an installed capacity of 6902 MW, of Hydro Power, currently, in Pakistan providing one-third of Pakistan's electricity needs. There is a potential of more than 50,000 MW of Hydro Power in Pakistan. Hydro Power development has suffered due to the KalaBagh dam controversy on which national consensus could not be developed. Sindh feared that Punjab would divert water depriving it with its share and that it required extra water to flow down Kotri to push the encroachment of sea into their lands. KPK feared that Nowshera would be inundated. Dams are not a zero-sum game. Dams increase water supplies and are an insurance against drought. Every country that has river, builds a dam. All the objections that have been raised against KalaBagh dam have been answered. Even after providing excess water flow down Kotri, there is excess water that can be stored.

Although water aspects are not the focus of this book or chapter, some factual discussion can be done year without causing major digression. In the adjoining table Pakistan Water Input-Output, we have provided data that indicates that even allowing for excess water flow into sea ,there is sufficient water available to build several dams. The total river flows are to the tune of 179 Km3, as against present consumption of 126 KM3, leaving an excess of 38KM3 that goes into the sea. It has been established now that only 11 KM3 of water is required for excess flow into the sea to prevent sea intrusion. And a net excess water to the tune of 27KM3 is available to build dams. The capacity of KBD is only 7.5 KM3.

It is the cheapest dam to build as it is close to plains and would have cost less than half of Bhasha dam. It would have irrigated a million acres in Punjab and a million acres barren land in Sindh. A lot of time has been wasted in controversy with no final output. KBD remains a contested project along with stalling of other opportunities. Only limited work has been done in this sector. Mangla dam height has been raised to increase its storage capacity. Now Mangla dam is the largest storage dam of Pakistan in place of Tarbela which storage capacity has gone down due to siltation.

Confidence building measures (CBM) are an important aspect of building dam in an environment where downstream provinces are skeptic of fair distribution of water. There are adequate technical solutions to record flow and consumption of water by all parties and provinces. In 2002, a water flow monitoring system was approved and later installed with the assistance of the World Bank with a financial outlay of 360 million Rupees. Reportedly, the system is not operational due to accuracy and operational issues. While provincial canal flow monitoring systems are working fine, national monitoring does not seem to work. The system was designed and built by a credible company, SIEMENS, there appear to be political issues involved. The system was supposedly installed as a confidence-building measure so that the availability of correct and reliable data assistants in removing doubts and suspicions regarding the actual eater usage by individual provinces. The system has been rejected and a new R&D based system under PCWR is under development under a modest funding. While one would want R&D projects to succeed, one would be sceptic of the outcome. Successful technical and political working of the water flow monitoring system is a must for future dam building projects, in particular, of KalaBagh dam. It is imperative that all problems and issues related with water flow monitoring be resolved earnestly.

Hydro development projects have suffered in the past due to KalaBagh dam controversy. Only lately some interest has been revived and many projects have been prepared for implementation. World Bank has agreed to finance DASU which is a 4400 MW project to be built in two stages of 2200 MW each. Bids are expected to be invited by 2018.Trabela dams power generation capacity is being upgraded through two projects; Tarbela IV and Tarbela V with capacity of 1410 MW and 1300 MW respectively. Tarbela IV is at an advanced stage of construction and may be completed and commissioned in 2018.Tarbela V financing has been approved and is also at its initial stages. Being extension projects do not require much time to complete as compared to the traditional hydro projects. Finally, Sukhi Kinari with an installed capacity of 900 MW has entered into implementation due to being taken under CPEC, perhaps a first or second private sector project in Hydro sector, it was mired in difficulties of arranging finances and achieve financial close.

Projects of 25000 MW capacity are at various stages of development. Mega projects are Bhasha (4500 MW) and Bunji (7100 MW).Bhasha is the most important project. It is a multi-purpose dam, producing electricity as well as storing water. There is no controversy around it, as there would be no canals in it and thus no provision of water diverting. Neither is there any issue of a city being inundated. Being free of internal issues, it is mired in external issues, as it is situated in what India calls a disputed territory. IFIs have asked Pakistan to get an NOC from India which is obviously quite unacceptable to Pakistan. USAID has shown interest but to the extent of scrutinizing the design which should not have been opposed, but a hue and cry was created in Pakistan by some circles and demands were made not to handover Bhasha design documents to USAID. Any technical input should be welcome, while we are free to do what we can, if we have the money.

Table 19: Hydro Power Potential in Pakistan

	Capacity (MW)	Under Implementation		Solicited Sites feasibility done	Raw Sites	Total
	In operation	Public Sector	IPPs			
KPK	3849	9482	2398	77	8930	24736
Gilgit Baltistan	133	11876	40	534	8542	21125
Punjab	1699	720	1028	3606	238	7291
AJK	1039	1231	3264	1	915	6450
Sindh				67	126	193
Balochistan				1		1
TOTAL	6720	23309	6262	4286	18751	59796

Source: PPIB

GoP has, however, continued with the preparatory operations such as land acquisition and others. More than Rs.100 Billion have been spent in the process making many locals quite rich. The project is ready for construction, but for financing. WAPDA has prepared a proposal under self and local financing arrangement according to which the project would have two parts or stages. The storage components would be built under public sector financing involving WAPDA's own income stream, PSDP funding and local bank borrowing. And in the second stage, Power component would be implemented under IPP regime. The proposal is quite reasonable and if nothing happens, we may have to go along these lines. Present government has vowed to start construction of Bhasha in the financial year 2017-18 which has already begun at the time of this writing (3rd July 2017).I am not sure what PSDP allocations have been made in this respect.

There is a parallel stream of negotiations that is going on with Chinese who have reportedly presented a very grand and ambitious plan of owning and operating the whole Indus cascade. It is a difficult issue and may involve bitter debate in political arena and thus may be time consuming. It may be desirable to separate Bhasha dam from the grand scheme for the time being and start its implementation with Chinese cooperation .They would be the most probable builders and EPC contractors in any case. The grand scheme can be discussed and negotiated and if consensus reached, Bhasha can be retrospectively included. Pakistan should not delay Bhasha and should start on its own and should not wait any further on it for want of agreement from any side, as we will narrate the case of Ethiopia which has built a similar project in similar circumstances.

Table 20: Hydro Power Projects under implementation or construction

Project Name	Capacity(MW)	Start yr	Completion
Neelum Jehlum	968	2011	2017-18
Tarbela IV	1410	2014	2017
Tarbela V	1300	2017	2019
DASU	4400	2016	2022
BHASHA	4500	2016	
MUNDA	800	2017	2022
SUKI KINARI	900	2016	2020
Golen Gol	108	2012	2018
Karot	720	2017	2022
Total	15106		

Source: WAPDA & PPIB

Problems and difficulties of Hydro Power

It is important here to narrate and discuss the risks and difficulties involved in Hydropower. It is not as wonderful as it appears to be in the popular circles. First of all, all hydropower plants do not have a water storage component. In fact except for a few, like Tarbela, Mangla, Bhasha and KalaBagh, no other hydro projects would provide any significant water storage at all. Apart from seasonal variations in water/electricity supply, there are long-term variations as well. Brazil depends on hydropower to the extent of 64-80%; it had severe problems of power supply due to drought conditions recently. Brazil is now taking steps towards having its access to more non-hydro sources. Due to climate change reasons, it has been projected that in Pakistan there would be large variations in precipitation and water supply resulting either in extraordinary floods or drought conditions. Having a large component of power supplies coming from hydro would compound our difficulties; no water and no power. Thus, there is an upper limit on the ratio of hydropower that one can have in the total energy mix. A large construction period, environmental issues and displacement of people have been the other reasons discouraging the introduction of hydropower in our energy mix.

One or two of the major fallacies based on which a popular support for hydropower has been built is that hydropower is cheap and that it brings water also. I have discussed the water component earlier; let us examine the cost issue. It is true that until recently, hydropower tariff was of the order of Re 1 per unit. Now, it is Rs 3.50 per unit, partly as a result of increase in Royalty payments to KPK. It is still an attractive price as compared to Rs 5-10 per unit of fossil-based power plant tariff. However, no more. All recent projects are to produce expensive hydroelectricity. Of all the people, KPK government is developing expensive projects costing in excess of 2.5 million USD per MW, resulting in a tariff of around Rs 10.00 per unit, approaching oil based electricity of today. Neelum-Jehlum has the same situation, although it had unique problems. There are issues of high relending charges which will result in a tariff of Rs 14 per unit. In any case, NJHPP tariff cannot come below Rs 10 per unit, whatever financial restructuring is done to it.

Table 21: Comparative Tariff and CAPEX of Various Hydro Power Plants

		KAROT	Suki-Kinari Mansehra	Gulpur-Mira	PATRIND
Location		Rawalpindi	-KPK	AJK	AJK
Capacity	MW	720	861	102	150
Annual Generation	GWh	3174		465	
EPC Cost	Mn USD	1277.78	1314.297	235.9	289.775
non-EPC Cost	Mn USD	409.62	388.46	80.9326	72.617
Total CAPEX	Mn USD	1687.4	1702.76	316.8326	362.392
CAPEX per MW	MnUSD/MW	2.34	1.9776	3.1062	2.4159
Debt Equity Ratio		80:20	75:25	75:25	75:25
Capacity Factor	%	50.83	40.4	53.30	
COGE(Tariff)	Usc/kWh	7.5746	8.8145	9.0241	8.2938
Expected Tariff Escalated		8.3321	9.6960		
ROE	%	17	17	17	17
Exchange Rate 1 USD=Rs	Rs	101.6	97.4	104.85	85
Owners		Three Gorges & IFC		Korea	
Tariff Year		Feb.2016(EPC)	2014(EPC)	Aug-15	Jan-12
Escalation		40% of CWC	do	Do	Do
Debt Rate		LIBOR+4.5%	LIBOR+4.5 %	LIBOR+5 %	LIBOR+ 4.75

Source: Compiled by the Author, NEPRA data

In China and Canada, hydroelectricity is priced at 4 cents, which is its average price internationally. In our neighborhood, in India, the CAPEX of hydropower is not so high. On the average, it is 1.5 million USD per MW, resulting in a tariff of about 5 cents per kWh, which appears to be quite reasonable. Far from home in Brazil, new hydropower projects are being auctioned with a price cap of 7.7 US cents per unit of tariff. Offers are expected around 6 cents. In Ethiopia, GERD dam is being built at a cost of 6400 Million USD, for 6000 MW and a very large water storage many times more than Tarbela. There is something terribly wrong or deficient in our hydropower sector; 2.5 million USD per MW vs 1.5 USD per MW elsewhere as mentioned earlier and a tariff of 10 cents plus vs 5-6 elsewhere. There is a difference of more than twice. I wouldn't jump to the usual reasoning, being very popular these days in political circles and otherwise. May be there are design and engineering issues, or of the monopoly of contractors coming from only one country, perhaps due to law and order situation. As there is long gestation period involved, there might be issues of escalation formulae which are normally understood with difficulty and the multi-currency issues as well. There could be contract management issues as well.

As billions are involved, it may be worthwhile spending some money to investigate the issue through credible international advice and a third-party input. Planning Commission had decided to commission a study on the issue of high CAPEX in hydropower in the context of NJHPP project. The outcome of the study should point out the reasons which would have general applicability as well. The study has not been commissioned yet, although, it is the requirement of ECNEC. I would not insinuate or blame unnecessarily. The study should be commissioned without losing any more time. Finally, it is in every body's interest that efforts be made to bring down the hydropower CAPEX so that more of hydropower capacity is brought into our energy mix subject to the issues that I have pointed out earlier.

Ethiopia builds a large dam out of its own resources?

Ethiopia has built a very large dam out of its own internal resources without resorting to borrowings from IFIs. The dam will be able to store 79 km3 of water and produce 6000 MW of electricity. For comparison sake, Tarbela had a storage of 13.69 KM3 which has come down to 7.993 km3 due to silting. The proposed Bhasha and KalaBagh dams are almost equal with 7.52-7.9 km3.Infact, as would be evident from the adjoining table, even if we build all the dams proposed, it would total to 42.3 km3 which would be only half of this dam.

Ethiopia is much poorer and smaller than Pakistan having a GDP 25% of Pakistan. It has suffered famines and civil wars. I have visited Ethiopia several times doing my professional consulting services for building parts of the power house locally in Ethiopia.

River Nile springs from its Lake Tana which flows through Egypt and ten other countries including Sudan and falls into the Mediterranean Sea. The erstwhile General Nasser built ASWAN dam on it. Egypt is largely dependent on Nile water and is worried over the consequence of the GERD dam, with a storage capacity as large that it may require 15years to fill to its full. Ethiopia faced almost the same problem as Pakistan is facing to construct GERD.IFIs and others wouldn't finance it due to potential and real objections and concerns of its big neighbor, as IFIs require NOC of India as a precondition for financing.

Ethiopia had no option but to rely on its own resources. It floated internal bonds and made deduction from salaries as loans and financed a 6.43 Billion USD project. It started construction in 1911 and is about to be completed and commissioned. We cannot do exactly what they did .But we need not. Our national and private banks have become quite big. Instead of financing sundry projects, they can be made to finance Bhasha. In fact, they have offered to do it to WAPDA which has proposed a phased strategy for going local. Chinese have offered to build it too. But they have proposed a very ambitious proposal to buy-out the whole Indus cascade which may attract a lot of concerns and in turn can cause delays. Let us go alone in financing. Eventually, others would join. It is hoped that GoP proceeds with the construction this year (2017-18) as promised last year.

Climate Change imperatives have made it essential that we start building as many storage dams as we can. Opposition to Kalabagh dam is eventually to go away in these circumstances. However, Bhasha is ready for implementation. It requires funding of the order of 15 billion USD, although over a period of 7 years. Top priority is to be given to Bhasha dam, unless we want to have a water and food crisis in next ten years. Bhasha must be our first priority now; beg, borrow or steal.

Table 22:Electric Power Installed Capacity vs Water Storage

	MW	Storage(Km3)	Current(Km3)
Tarbela	3478	13.69	7.933
Mangla	1000	9.12	9.12
Chashma	184	1.07	0.429
DASU	4320	1.41	
Bhasha	4500	7.9	
KalaBagh	3600	7.52	
Munda	740	1.59	
Total Storage Potential	17822	42.3	17.482
GERD Ethiopia	6000	79	

Source: WAPDA, GERD

Table 23: Pakistan Water Input-output Table

	MAF	KM3
Total River Flow	145	178.785
System Losses	12	14.796
Flow into the sea	31	38.223
Min. Flow against sea intrusion	8.86	10.924
Current Storage	14.15	17.447
Desired Storage	22	27.126
Storable Water	22.14	27.299
Water consumption	102	125.766

Source: WAPDA & Water info

14. Hydro Royalty

Hydro Royalty or Net Hydro Profit (NHP) as it is called in our constitutional (1973) terms, has been a major dividing and polarizing issue. Some provincial leaders of KPK have issued extremist and incendiary views on the subject demanding early resolution and implementation of their demand in this respect. CCI has reached a consensus on it and its decision has been passed on to the relevant authorities. WAPDA has applied for its tariff to NEPRA containing the approved figures and a public hearing has been scheduled for its consideration and ultimate approval.

CCI has reached a consensus on Rs.1.10 per kWh as NHP settling many issues of payables and receivables. Until now, WAPDA has been paying Rs 0.58 per unit as NHP or royalty. Actually, the real and internationally recognized term is royalty, as is royalty issue dealt with in the case of other resources such as minerals, Oil and Gas etc. There is no issue there now on the latter, but the term NHP (Net Hydro Profit) has created difficulties. It is difficult to define and exactly determine NHP as there could be many interpretations. The late Aftab Ghulam Nabi Kazi (AGNK), a towering personality of his time in government bureaucracy, did one of the interpretations.

AGNK formula defined NHP (putting it in nut shell and simply) as the difference between average cost of production of electricity (average whole sale prices) and the production cost of hydroelectricity. Some people argued that instead of average COGE (Cost of Generation), it should have been the next best rate from which the difference should have been calculated as NHP. AGKN formula created quite some controversy as it yielded a NHP of Rs 1.80 per unit at a time when the hydro tariff was of the order of Rs.0.50/kWh or even less. Another reason for controversy has been the unduly low COGE of Tarbela due to the depreciated assets having no or very little book value. A tribunal was appointed which gave its ruling in favour of AGKN formula. One of the tribunal members while signing the tribunal decision also said that let us not kill the goose that lays eggs. The goose (WAPDA) has since been protesting and asked GoP to foot the bill as WAPDA did not have the resources to pay such a high royalty/NHP.

KPK people and politicians should feel happy and satisfied. They have got the highest possible award of NHP/Royalty, if compared with other countries of the world, as would be readily seen from the adjoining table. It is more than twice than that of India and several times higher than in other enlightened countries known for internal fairness and equity policies.

A lot of water has flown under the bridge since AGNK did this exercise. A lot of national and international data is available to make a reasonable determination today than it was possible in earlier days. There is a need to reexamine the issue on a permanent basis. Current CCI decision, although a welcome one creating consensus, does not lay down a formula. It is a figure that may be contested later due to inflation and other factors.

Interestingly enough NHP as defined by AGKN renders hydropower as expensive as any other resource for if there is a difference, it would be charged as NHP. Thus no more slogans of cheap hydro should have been raised since NHP concept emerged.

Hydropower is getting expensive by the day. All new projects are yielding a COGE of Rs.8.00 per unit or even more while at the same time solar and other renewable energy prices are coming down to Rs. 4-6 per unit. Local gas based electricity is already available at a COGE of Rs. 5-6 per unit. Low oil prices have caused reduction in average tariff. When furnace oil is eliminated from the energy mix and more economic resources come in, the situation may drastically change. There would be no NHP. But KPK would need revenue and it should continue to get revenue. This indicates the need for reconsideration of the NHP issue and to settle it on a more sustainable basis. It does not mean, as should be evident from the aforementioned, that I mean to oppose the current determination of NHP at Rs.1.10 per unit. It is a consensus creating decision and it should be welcomed.

Royalty is generally considered a reward to the people and governments of a region or country whose resources are consumed (many countries do not recognize royalty on hydro as it is a renewable resource) and regional and local environment is usually compromised causing land-use changes, pollution and general disturbance, unpredicted changes in life and power structure etc. Many royalty provisions therefore require distribution of a share to local people and government. In my opinion, hydro royalty should have been spent on people in social sector and improvements in the areas, which have been affected by resource development activities. When Tarbela was developed, compensation rules were not fair. Although, today more than 100 billion Rupees have

Table 21 Comparative Royalty Rates in Hydro sector in Various Countries

Country	$/MWh	Rs/kWh (Pk)	Comments
China	0.75-1.25 USD	0.075-0.125	
Brazil	1.8 USD	0.18	6.7% of Sales Price
Saskatchewan (Canada)	5.1 Canadian $	0.3825	Canada Hydro Electricity price 40 Can$/MWh
Quebec	3.82 Canadian $	0.2865	
Ontario	3.80 Canadian $	0.2865	
Manitoba	1.51-3.11 C$	0.013-2.33	
USA	2.01 USD	0.21	
India	4.8 USD	0.48	12% free electricity; electricity price 4 cents
Nepal			2% during first 10-12 yrs and 15% thereafter
Pakistan		1.1	100Rs=1 USD

Source: 1)Hydro Power Royalties: A comparative Analysis of major producing countries by Pierre-Olivier Pineu et al,MDPI,20th April,2017; 2)CERC India

been spent on land purchases and acquisitions. KPK government has instead opted to develop hydro sources from a Hydro Development Fund that has been created to channel Royalty income. KPK has grinding poverty and much needed social sector assistance is required there. KPK has no obligation to generate or provide electricity. It should not divert its much-needed resources to something for which Federal government is responsible.

Water User Charges from IPPs

While NHP is payable by public sector projects to the provincial governments where the project is located, IPPs under the current dispensation are required to pay user charges. It is due to the constitutional aberration that NHP has been required to be paid by the government projects only. Hydro Royalty, whether termed as NHP or water user charge ought to be payable irrespective of the ownership being in public or private sector. It is compensation or user charge whatever may be the term adopted. Currently, IPPs are liable to pay a water use charge, which is Rs.0.45 per kWh. Until now, it did not matter as there was no hydropower project in the private sector. However, in future, many hydro power projects are expected, a number of those are in pipeline already. It is only fair that the IPPs be required to pay as much as public sector pays, irrespective of whatever the payment is called; water user charge or NHP or Royalty.

Royalty Payments to AJK

Currently, no royalty is paid for Mangla dam, which is situated in AJK. There are, supposedly, some political or constitutional implications or impediments in applying NHP provisions to AJK.I am not sure what are those. There shouldn't be a problem in paying, there can be issues on extraction or levying charges. This is an anomaly that should be removed. Some Kashmiri nationalists raise it as a negative point against GoP. It should be, however, noted that AJK gets payments on many accounts and also there is almost a free electricity regime in AJK. The electricity tariff there is low and even that is not paid and theft is rampant which is taken lightly by WAPDA. In order that accountability is created, accounting should be right and subsidies or levies should be accounted for in a regular manner. It would be therefore, only fair that Royalty payments be initiated on Mangla and upcoming Neelum- Jehlum and other new projects.

Hydro Royalty to Punjab

Up to now, successive Punjab governments did not want to charge royalty. Perhaps, there was a political stigma to it being an unnecessary extraction. No royalty has been paid on Ghazi Barotha. There is some dispute as to where the project lies. Apparently, this has been resolved, and one does not hear much about it. Punjab government has demanded royalty on the lines of KPK and CCI has ascended to their demand. Even back payments are being made. Therefore WAPDA has included Punjab Royalty payments in their Tariff application. So now that Punjab is getting hydro-royalty, AJK should also get it. However, there may not be a case for small hydro projects of Punjab to get NHP, as its cost of generation is very high and there is no NHP or profit to be charged on the power producer in that case, although it is a small amount.

Concluding, although consensus has been reached on numbers, which is to be welcome, it is temporary. There is a need for building a lasting and sustainable solution based on definite definition and methodology on which there should be consensus. It should not be too less for the recipient and too much for the payer and producer and should be comparable with other jurisdictions. Some political accommodation would have to be reached on constitutional niceties as well. It will be a tall order, but can be done.

PART III: OIL & GAS

15. Restructuring Oil and Gas Sector

Oil and Gas (O&G) sector has defied reforms and restructuring for a long time now, despite demand by the stakeholders and recommendations of others. Even 18th Amendment could not move them. It is a big joy lording over such a big and in many ways quite a prosperous sector. So much so that recently, some wise men of MPNR proposed investments and activities in power sector as well.

Over the years, the power sector has undergone major restructuring from a monolith of WAPDA to the DISCOs and GENCOS. More reforms are in continuous deliberations. On the other hand, the O&G sector has not progressed as is indicated by deteriorating performance of the gas companies, depleting gas reserves, and underdeveloped transmission infrastructure. Exploration companies have been complaining all along of bureaucracy and red tape in petroleum concessions. Fortunately, Minister Khaqan Abbasi has turned his attention to it and is in favor of a fast track change. World Bank consultants have been appointed who are in the process of consultation and finalization of their recommendations. The purpose of this article is to inform the stakeholders of the issues involved.

There has been no dearth of advice. I made my submissions in this respect in my books (Pakistan's Energy Development (2009); and Issues in Energy Policy (2011). There are others who made recommendations in this respect. There is almost a consensus among the policy experts that reforms should have the following components (in fact this mantra is common for all utilities including electricity as well;

1) Privatization

2) Separation of Transmission and Distribution

2) Decentralization into smaller distribution companies

3) Central System Operator, controlling transmission, if not owning the infrastructure

4) Separation of pipes from supply and selling business

5) Competition in retailing (which I think will be a bad decision as we shall see later in this discussion)

6) Independent regulator (which perhaps is the only step taken, irrespective of the independence issue)

7) Role of provinces

Reverse has happened with respect to privatization, although the story is rather old. There used to be Karachi Gas Company in private sector - very efficient and profitable with almost no losses - which was nationalized in 1973 as part of larger nationalization campaign. Issues of restructuring and provincial role have thwarted later efforts of privatization. Current thinking is, and perhaps rightly so, that restructuring should precede privatization, or else, there would be legal constraints. Moreover, privatization of the whole gas sector in the present form may be an unmanageable transaction.

Based on the above-mentioned principles, the consensus is emerging on establishment of smaller gas distribution companies aka DISCOs; 3 distribution companies for Punjab (quietly considering political aspirations of Southern Punjab); two or three in KPK, Hazara division, Central KP and Kohat-DI Khan); two or three companies in Sindh and one or two in Balochistan. Provincial lines should not be crossed for an eventual provincial role, as we shall see later.

There can be a difference of opinion in the case of transmission. Some provincial enthusiasts argue for provincialization of transmission. It would be a major disintegration benefiting no one, as the transmission is a common facility. KPK's access to gas would depend on transmission from Sindh etc. Thus it may be a good idea to make one national transmission company by merging SNGPL and SSGC. The transmission company may, in initial days, be in the public sector and may eventually be privatized except the supply company or it may be that the gas distribution companies buy gas from the gas producer or importers (RLNG) directly paying the wheeling charges to the transmission company. The advantage of separating commodity/gas selling from the transmission is that competition and investments may come in the pipeline sector as well. Perhaps too utopian in our circumstances. In a society where power, even brute power, reigns higher than competence and efficiency, and may still remain so, for quite a while, trying to approximate pure competition by creating smaller entities may be counterproductive. Some balance is to be maintained among the extremes: one or two public sector companies and the 30-40 companies that may emerge under utopia. Utopia (of equality and public ownership) failed in the Soviet Union and International trade Utopia is being challenged by President Trump.

Another issue is whether Gas DISCOs should be both buying and selling the gas or should only provide distribution service. Gas importers/producers/ supplier should engage in retailing also. The simpler arrangement would be to have Gas DISCOs operate on the lines of electricity distribution companies whereby large consumers may have the option to buy directly from suppliers and pay wheeling charges to transmission and distribution companies.

The role of provinces is a major issue, which complicates and obstructs decision-making almost at every stage, especially after 18th Amendment. Gas is provincial subject under the 18th Amendment. During approval processes of RLNG based power projects in Punjab, hair-splitting arguments emerged from Sindh demanding role in RLNG in addition to Gas, as to them constitution does not differentiate between RLNG and Gas. It mentions of Gas only. I could never understand the logic of provincial role in imported LNG. Provincial eminent domain has been recognized in terms of land resource argument. I am not aware as to the outcome of that controversy.

Provinces get royalty on locally produced gas and share in Excise duty and GDS. The residual issue is of control. However, when gas comes into the pipe, there is no distinction as to which molecule is local and which is not (Molecule labeling technology is being developed in the context of tracing material origins for international security and safety purposes). Separate control and management of molecules would still be impossible).

There may be a case for direct provincial role for oil and gas exploration. Classically, DGPC used to be under federal government without any provincial representation. Lately, provincial representation has been provided in DGPC. There have been complaints by the foreign and local investors against DGPC of red-tapism, inefficiency, and confusion. Two-stage concession processing has been initiated by KPK where exploration activities have been put on fast track by the appointment of some very efficient persons heading the relevant departments and companies. There is an argument for provincialization of DGPC as it may promote competition and distribute power. There may be capacity issue for Balochistan, for which some solution may be evolved.

The provincial role may be included in Gas Discos by providing for provincial shares in its capital structure in the range of 26-49% varying with individual situations of affordability. In Central Transmission companies, perhaps there is no case for provincial role. I have long been proposing to consider double-board system in selected companies and institutions. This system has been largely adopted in public listed large companies, albeit in a different context. An oversight/supervisory board (without or limited financial and management powers) may be introduced not in utilities only which is a subject of discussion here, but may be extended to other areas of service providers.

An RLNG supply company (Pakistan LNG Ltd) is in the initial formation process following the tenet of separating pipe from commodity. Perhaps, the intention is not making a public sector monopoly. A role for RLNG imports in the private sector has been provided. CNG sector has been given permission to import LNG for its members, although the sector could not organize itself yet into doing so. Hopefully, others like IPPs would be able to import for themselves or buy from local producers directly. There has to be a company or entity in public sector, at least initially, whose responsibility should be to take care of supply and demand balance and to see to it that no shortages occur. In my opinion, this company should be made responsible for all gas supplies ala CPPA for power sector, if not immediately but eventually, a form of a supply integrator, some kind of PSO for Gas. ISGS had initially been formed for similar purposes initially for handling Iran-Pakistan Pipeline project and has also been responsible for TAPI.ISGS is now also handling RLNG terminal and Gawadar-Nawabshah pipeline. A company named Pakistan LNG terminals Ltd is also being organized which I think would be too much of a hodgepodge. Pakistan LNG Ltd to be eventually responsible for overall gas supply may be entrusted with the task of controlling the RLNG terminals also.

Establishment or introduction of gas market at wholesale level is also a matter worth consideration. With the advent of RLNG, the issue has become more important. A certain percentage of market could be reserved for market operations, initially, for introductory and training purposes. LNG imports could be on free list. Importers should be able to sell directly to large consumers at unregulated prices. This may enable price discovery. PSO is importing petroleum products along with all others who are importing petroleum like refineries and IPPs. The open economy under market principles, as much out of the grip of politicians and bureaucracy, promotes growth and efficiency and lowers prices. However, this is a rather complicated subject where in chances of market manipulations are always there for which adequate oversight infrastructure is required as well. Some beginning, however, may have to be made in this respect.

All of the above is a tall order. It cannot be possibly done in one go. Some kind of phasing may have to be done. But the risk in phasing is that much may not be done. Reform and restructuring is on table for a long time now. Elections are to be held in 2018.Legitimacy and political authority to take such major decisions wither away even earlier. The restructuring would be much easier while all the companies are in public sector. Divesting of distribution assets from SNGPL and SSGC and bringing about GASDISCOs as discussed earlier can be done in the tenure of this government. After this restructuring is completed, the next government can undertake privatization and undertake residual reforms. Indeed, virtual restructuring can be undertaken immediately through cost accounting. A thousand miles journey starts with the first step. Let me remind the readers that before Margret Thatcher's reform period, it used to take months even a year in the UK, to buy a gas stove from British Gas, a government monopoly. We would be starting from a much better condition. Gas stoves and heaters have always been available here from eager shopkeepers. Consultant's report should be coming in. Let us start the work earnestly.

16. Natural Gas and LNG Demand

Natural gas has emerged as attractive and popular energy source. Firstly, oil and gas prices have come down. Natural Gas has traditionally been sold at 75-80% of the oil price. It is expected that this ratio would sustain. Secondly, LNG prices have also come down and availability has increased. And thirdly, Combine Cycle Plants' efficiency has increased phenomenally up to almost 60%, almost twice or more the present level of average efficiency. Due to its relatively clean burning characteristics and lower carbon footprint, it is also being preferred by many countries.

We are passing through a general energy crisis, of both gas and electricity. Apparently, there should be no reason for joint occurring of both. Lack of electricity does not cause reduction in gas production, although reverse may be true. Major expansion has been launched in Pakistan. 5 LNG plants are in pipeline with a total expected capacity of 2-3 bcfd (400-500 mmcfd gasification capacity per plant).

Gas production has stagnated for almost a decade since 2008 at around 4 bcfd. It was preceded by a growth period of 2002-2007, in which gas production increased from 2.5 to 3.8 bcfd. Ironically, reverse has been the case of oil; in almost the same growth period (2002-11) of gas, oil production stagnated at around 65,000 barrels per day. Thereafter, Oil production started increasing rather phenomenally and reached a production level of around 100,000 barrels per day in just 5 years. Why production in one has caused stagnation in the other? I have heard, the exploration effort is common; it is almost a chance that either oil or gas is discovered. Oil is useful, even if there is no oil refinery in the country or nearby. It can be exported using trucks and ships. But gas is useless unless there is a gas processing plant nearby or a new one is installed. It takes around seven years to explore and develop a gas field, although oil may require less. Hence, it is important to develop a reasonable forecast. Unfortunately, I have not come across a reasonable study on the subject unlike electricity where several major studies have been done which we have referred to elsewhere in the electricity section.

We will undertake here a rough-cut analysis utilizing growth rate methodology. Pakistan's growth rate of economy has been around 5% p.a. in 30 out of last 42 years. We will also make an assumption that gas demand or consumption would grow at the same rate as that of the economy. However, gas consumption (not demand) has stagnated since 2008 and has even decreased. We will, therefore, consider the growth rate in the period, when there was no shortage and assume that in a period when gas supply improves, growth rates would recover to their previous levels.

Table 22 Local Oil and Gas Production Trend (2002-15)

Year	Oil (bbl/d)	RoG(%)	N.Gas (mmcfd)	RoG(%)	Oil to Gas Ratio
2001-2	63,547		2,531		0
2002-3	64,269	1.14	2,719	7.43	0.14
2003-4	61,817	(3.86)	3,286	20.85	0.11
2004-5	66,079	6.71	3,685	12.14	0.10
2005-6	65,578	(0.79)	3,836	4.10	0.10
2006-7	67,438	2.93	3,873	0.97	0.10
2007-8	69,954	3.96	3,973	2.58	0.10
2008-9	65,844	(6.47)	4,002	0.73	0.10
2009-10	64,948	(1.41)	4,063	1.52	0.09
2010-11	65,866	1.45	4,032	(0.76)	0.09
2011-12	67,141	2.01	4,259	5.63	0.09
2012-13	76,277	14.38	4,126	(3.12)	0.11
2013-14	86,534	16.14	4,092	(0.82)	0.12
2014-15	94,493	12.53	4,016	(1.86)	0.14

Source: HDIP 2015

We have divided the end-user sectors into two classes: one is the non-policy sectors consisting of the following: a) domestic; b) commercial and c) general industries and the other class is of power, CNG, fertilizer and cement etc. These sectors can pass on their costs to others or can afford it otherwise or may enjoy government subsidy. Cement sector has almost shifted to coal, and Pakistan Steel has been closed down. It is assumed, consistent with the policy proposals that are circulating, that policy class of end-users would be given LNG and the non-Policy class (domestic, commercial and general industries) would get the cheaper locally produced gas. Thus this would enable us to forecast the demand of Natural gas (local) and imported LNG.

Figure 12 Local Oil and Gas Production Trend (2002-15)

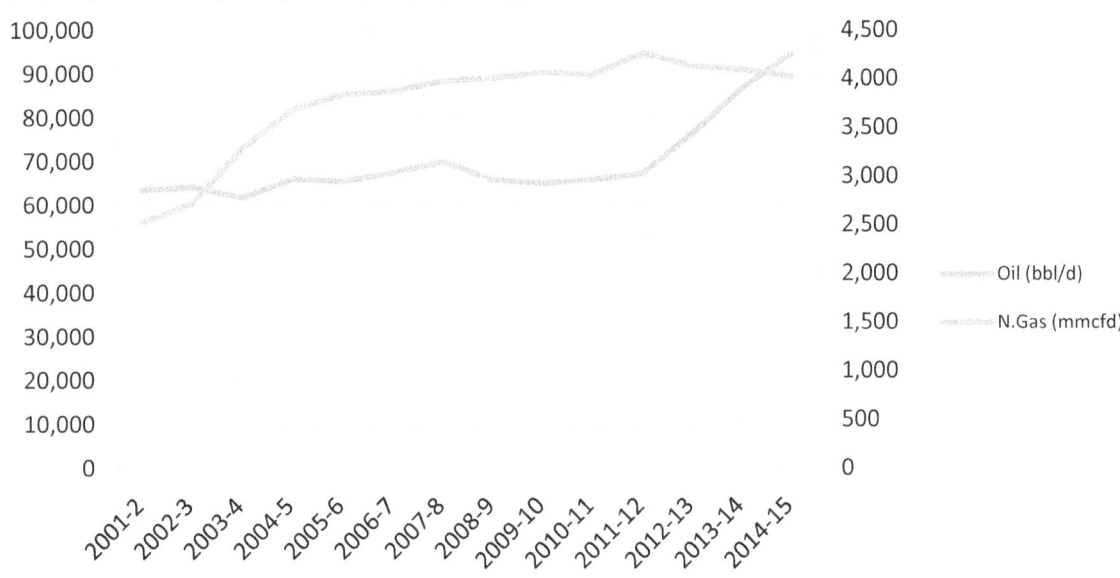

The adjoining Table 21 provides the historical consumption data for the years 2002, 2008 and 2015 and growth rates for the periods (2002-2008) and (2009-15). The base year has been taken to be 2002-2003. Forecast has been developed for the year 2015 and 2025. Calculations for 2015 will provide us a mechanism to judge the compatibility of our numbers with the reality. Following may be noted:

1) The share of Policy and Non-policy group/class is roughly equal, 50% each, and it broadly remains almost the same in the forecast years and period.
2) Three Forecasts have been developed(Table:23) ,of which we have adopted Forecast-I, for the reasons explained in the following:

2.1) Under Forecast-I, the rates of growth have been taken as historical which come out to be a composite of 5% and 6%. The predicted forecast under Forecast-I of FY 2015 is 1.786 TCF (4.895 bcfd) against the actual of 1.22 TCF (3.356 bcfd) giving a shortfall of 0.566 TCF (1.539 bcfd). The predicted forecast for 2025 is 2.928 TCF (8.0 bcfd) as against the actual of 1.22 TCF (3.356) of 2015, giving a shortfall of 1.708 TCF (4.644 bcfd)

2.2) Under Forecast-II, the rate of growth is taken at 6% p.a. It results in, what we think, an excessive forecast of 9.4 mmcfd, leading us to reject it.

2.3) Under Forecast-III, assuming a rate of growth of 5% p.a., we get a projection of 7.426 mmcfd for the year 2025, which is close to Forecast-I. Thus we select Forecast-I to be the acceptable one and adopt it for policy discussion.

3) It can also be said that RLNG demand for the year 2025 would be around 4 bcfd, assuming that the local production is maintained at 4bcfd. If local reserves fall, LNG demand in 2025 would be higher .It would increase at a rate of 200 mmcfd in the intervening period creating another shortfall of 1.4-1.6 TCF (4 bcfd) and the gap to be filled by RLNG.

4) With a total of 8 bcfd demand, power sector allocation would be 985,320 mmcf per annum (projected as in Table 22), which would be able to power 16,000 MW of Steam power plants with an average efficiency of 30%. Existing capacity of oil and gas fired power plants is 14,630 MW. Assuming all are new NGCC power plant with, 57-60% efficiency, almost 30,000 MW could be produced. However, there would be a mix of both technologies, bringing the number in between depending on the mix. 3,600 MW of RLNG plants are already under installation. Three more RLNG plants are being actively talked about: one in KPK on its own gas (or LNG), another in Muzaffargarh eventually replacing the GENCO, and last one in Karachi.

Table 23 Natural Gas Consumption (2002-2015)

	2008 (mmcfa)	RoG(%) 2003-8	Share(%)	2015 (mmcfa)	RoG(%) 2010-2015	Share(%)
Non-Policy Sectors						
Domestic	204,035	5.9	16	278,069	4.9	22.7
Commercial	33,905	8.3	2.66	35,187	-1.01	2.87
Gen. Industries	305,662	15	23.97	239,591	-5.7	19.56
Sub-Total	543,602		42.63	552,847		45.13
Policy Sectors			0	7,623	-10.2	0.62

Pak Steel	16,901	5.6	1.33	7,623	-10.2	0.62
Cement	12,736	29.9	1	831	-15.6	0.07
Fertilizer (Feed Stock)	160,062	2.6	12.55	166,903	-1.01	13.63
Fertilizer (Fuel)	40,001	0.2	3.14	58,609	5.7	4.78
Power	429,892	5.1	33.71	371,562	0.3	30.33
CNG	72,018	44.8	5.65	66,517	-7.6	5.43
Sub Total- Policy sector	731,610		57.37	672,045		54.87
Total	1,275,212	7.9	100	1,224,892	-0.8	100
Total (per day)	3,494		0.26	3,356		

Source: HDIP 2016

Eventually, and not far, all GENCOs would be closed down, which would create a market justification of another 4,000 MW NGCC. This may happen in the period between now and 2025. Already, it is alleged that the existing capacity is not being utilized due to expensive oil and lack of gas. Some IPPs would also be closed down in the period. HUBCO (900 MW) is already a candidate, and there are several others. One can visualize about 10,000 MW of high efficiency new NGCC power plants in the period leading to 2025 and another 10,000 MW in the period after 2025 replacing all the existing oil and gas power plants whether GENCOS or IPPs.

There is not more than 20% price advantage between Furnace Oil and RLNG on existing low efficiency steam plants. RLNG, twice as expensive as local gas - the prices of latter poised and proposed to increase at-least for power plants, the real advantage and cost rationalization would be obtained through high efficiency NGCC. Americans would be happy and would hopefully stop quiet criticism, if not opposition, to CPEC, as there would be a balancing effect. GE and Alstom, the latter having been bought by GE, are supplying NGCC power plants.

Table 24 Natural Gas Consumption Forecast (2015-2025)

	Consumption (mmcf) 2008	Assumed ROG (%)	Forecast Cons. (mmcf) 2015	Forecast Shortfall (mmcf) 2015	Forecast Cons. (mmcf) 2025
Non-Policy Sectors					
Domestic	204,035	6	306,793	28,724.20	549,420
Commercial	33,905	6	50,981	15,793.58	91,298
Gen. Industries	305,662	6	459,603	220,011.63	823,078
Sub-Total	543,602		817,376	264,529.42	1,463,797
Policy Sectors					
Pak Steel	16,901	0	16,901	9,278.00	16,901
Cement	12,736	0	12,736	11,905.00	12,736
Fertilizer (Feed Stock)	160,062	3	196,856	29,953.07	264,558
Fertilizer (Fuel)	40,001	3	49,196	-9,412.82	66,116
Power	429,892	5	604,901	233,339.21	985,320
CNG	72,018	3	88,573	22,056.06	119,035
Sub Total-Policy sector	731,610		969,164	297,118.53	1,464,666
Total (mmcfa)	1,275,212		1,786,540	561,648	2,928,462
Forecast-I Total per day(mmcfd)	3,494		4,895	1,539	8,023
Forecast-II		6	1,917,447	692,555	3,433,856
Forecast-II (per day)			5,253	1,897	9,408
Forecast III (per day)@5%					7426

Source: compiled by the author

5) There will be, however, competitiveness issues that have been examined elsewhere in this book. According to NEPRA's April 2017 DISCO fuel charge schedule, the variable (fuel charges) per kWh for DISCOs are the following: Gas, Rs 4.2919; FO, Rs.10.24 and RLNG, Rs8.08.One has to add at-least Rs. 2.0 per unit to this figure to arrive at total generation cost (COGE). We are entering a low price regime. Coal price is expected to come down from its current high level, which appears to be a temporary and transient

issue due to crisis in Indonesia. Thar coal is also expected to be available at a lower price on expansion. Thus coal electricity may be available at lower than the existing tariff of Rs. 8-8.50. Solar power has come down to Rs. 5.25 per unit (Zorlu proposal for QA Solar Park) and may come down further. National and international pressure to induct solar and wind will mount. Hence, there would be downwards pressure on RLNG, also because it is an imported fuel creating foreign exchange liability. Local gas, hopefully, would also be produced eventually. Thus the forecast gas demand of 8 bcfd, double than that of today's consumption, may not be realized.

6) RLNG investments in Regasification terminals over and above, the existing target of 5 RLNG plants with a combined capacity of 2 bcfd, may not be forthcoming. We have argued elsewhere for alternative RLNG approaches (technical and financial, such as debt financing instead of current lease financing, and other technical approaches). Such changes, if implemented, may facilitate higher induction of RLNG infrastructure. Hence, the kind of projections that have been discussed may not be achieved under an RLNG regime, but are quite possible under, hopefully, cheaper local gas.

7)Here we would like to undertake a little discussion on RLNG tariff issued by NEPRA in 2015.It has issued two extreme tariffs (Are we extremists as a nation?). There is a capacity factor of either 60% or 92 %; 60% being too less as compared to NGCC plants already operating in the country at 80% and 92% being too high to be impossible to achieve. Obviously, it was done under pressure from the promoters (MoWP) being eager to show the good economics of the project for the time being. Availability factors are being mixed with capacity factor. A plant may be available 92 % of the 8760 hours in a year, but it may and will not be utilized all the time it is available. Nuclear power plants, even, do not have such a high capacity factor. A more reasonable figure would have been 80%.

8) Moreover, tariff has been calculated at 10, 11 and 12 USD per MMBtu. We will have to transform these figures, for the sake of comparability, to 8 USD per MMBtu RLNG rates and a capacity figure of 80%. We have taken the data for 800 MW project capacity at foreign financing. The fuel charge comes out to be Rs .7.312 per kWh, which appears to be too high than Rs 8.00 per unit fuel charge of existing low efficiency NGCC plants. The capacity charge would be Rs.2.50 per kWh at 80%. Adding an O&M of Rs. 0.50 per kWh, the total COGE should come out to be Rs 10.312 per unit. Possibly lower efficiency figures have been assumed by NEPRA. This may be attractive vis-à-vis furnace oil but would not cut much ice against other competitors.

9) Simple Cycle efficiency today stands at 40%. As early as 2010, even F-class (early generation) turbines achieved efficiency levels of 59% and the H-class (new series) turbines are giving 60% plus in combined cycle mode. Utility level efficiencies are a bit lower due to a variety of reasons. However, to be realistic one could assume 60% as a viable efficiency figure for new Tariff purposes. With an efficiency of 60% and an RLNG price of 8 USD per MMBtu, the fuel cost would come out to be RS 4.00 per unit, just half of the existing low thermal efficiency plants. Adding CAPEX of RS. 2.50 and OPEX of 0.50, the COGE would come out to be Rs 8.00.

10) Similarly, with an allocation (as predicted in table 22)264,558 mmcf per year in 2025,a production level of 7.35 million tons of Urea production will be sustained(assuming a natural gas consumption rate of 36 MMBtu per ton).

The general statements coming out of MPNR are compatible. MPNR says that current consumption is about 4 bcfd per day with a shortfall of 2 bcfd, which will in near future will increase to 4 bcfd, if LNG is not imported. While it is good to plan and implement imported RLNG projects in the intervening period, it is imperative that efforts for local gas production should be enhanced. A new gas policy is due with pricing rules updated to reflect current realities. When the companies were asking for 10-12 USD per MMBtu, we were giving 6 USD. And now, it is 3 USD per MMBtu against the 10 USD per MMBtu of LNG. Local production should be more beneficial and thus valuable than imported LNG. However, the issue is not so simple. On the other hand, the prospects of increasing local production at the prevailing prices are not very bright. Foreign companies cannot be attracted. National companies can continue. But they have to be energized if not incentivized. Local production resources and infrastructure such as EPC contractors, drillers, service companies etc. would have to be created and a policy brought out to achieve this. It is not a difficult task.

Gas Demand Curtailment Options and Prospects

There may not be much possibility of demand curtailment in non-policy sector comprising domestic, commercial and general Industries. Any curtailment there hurts the economy, as there are no substitutes to gas in these end user sectors. The opportunities for demand curtailment may only be there in policy sector but even in that fertilizer sector demand can only be curtailed, if there is a shift to coal gasification based fertilizer production. Existing urea plants can be modified to run on coal gasification plants to be sited in Thar. At least 2 or 3 plants which are located near Thar may be attempted. This may become essential, if there are gas shortages due to reduced local production or possibly extra-ordinary price increase, as has happened earlier.

It is only power sector, where major curtailment may be feasible. GENCOs can be closed down in the period leading to 2025 being highly inefficient and irregular. In the period, 2025-2035, all existing IPPs running on gas and furnace oil may be closed down. The 16,000 MW can be substituted by 8,000 MW of RLNG capacity, which is twice more energy efficient. In that way, power sector demand can be halved from the existing level. From this discussion, it appears that there is no option but to increase local production or fall into the trap of import dependence (mostly LNG) and the consequent adverse impact on the economy.

Projected Demand-Supply Outlook-2025

Let us now put together all the numbers that we have generated or tabulated. It appears that the demand projection of 8.0 bcfd appears to be a dependable number being the middle of the two extreme demand projections of 7 or 9 bcfd. It also agrees with OGRA projections. On the supply side, there are two scenarios: Scenario-I, rather optimist scenario which assumes that local production would remain at the current level, which in turn means that local resources would be developed and further gas fields would be found and developed. This also assumes that IP and TAPI are implemented. Under this scenario, practically, there may not be any shortage. Under the second scenario-II, the situation appears to be quite bleak and large shortages may emerge. Local production goes down, and IP and TAPI do not come up, then, the gap would be of order of 2.2 bcfd, almost unmanageable causing large stresses and economic disorder. Dependence would emerge on RLNG altogether. Although, there is not much of a difference between IP, TAPI and LNG are all equivalent in terms of pricing.

An important conclusion that emerges is that local production should be enhanced. It would be cheaper and save foreign exchange. RLNG should not be grudged as some people do, it is a short to long term solution, requiring least of investments. All that is required is a terminal which costs around 300 million USD. Regasification tariff can be brought down as has been discussed elsewhere almost half that of Engro. It has already come down in the second terminal. Engro always manages to make unethical profits. In case of Thar coal, they lobbied and managed to get an unreasonable return of a 20% IRR which would make Thar uncompetitive. And let us hope that Iran-Pakistan pipeline and TAPI are finally implemented, although the political circumstances indicate otherwise, which means that LNG may have to be augmented along with extra-ordinary efforts to enhance local production. It is not impossible. Gas resources are there, these have to be explored and developed.

There is yet another point of view and a serious one. Many people argue, that dependence on imported fuel is not a good policy and that a gas demand of 8 bcfd by 2025 is not sustainable. It would grow further and that large scale imports are not feasible. This practically means a gasless scenario, meaning thereby no expansion of gas based facilities. This would also entail retirement of GENCOs, once new generation capacity is commissioned and comes on line and eventual retirement of thermal IPPs. This is already happening. Capacity utilization of IPPs is under 50%, as we have indicated earlier due to unavailability of gas and furnace oil. This would mean an almost ban on imported fuels based power plants including NGCC and coal power plants in Punjab under CPEC, as MoWP has already suggested but has been opposed by powerful interests. In a falling exports scenario and not much hope for a significant increase in foreign exchange earnings, phenomenal import bill should not be created through imported gas and RLNG.

Table 25 Natural Gas: Projected Demand and Supply (mmcfd) 2025

	Scenario-1	Scenario-2
Demand	8000	8000
Local Production	3500	1864
IP	750	750
TAPI	1350	1350
LNG	1800	1800
Supply	7400	5764
GAP	600	2236

Compiled by Author using various national and international sources

It is a good interim solution, which should not be converted into a long-term liability under a naïve enthusiasm. This would mean more investments in solar, wind, hydro and Thar coal. Natural Gas resources to be conserved and would be allowed for domestic and may be industries under priority till all local natural gas resources are exhausted or new ones discovered. Whatever little gas imports are feasible would go to the priority sector. Most fertilizer could be produced through coal. There may not be a choice but may be a dictate of the circumstances. If MPNR cannot guarantee the local production of gas, it should not lobby for gas based energy regime.

17. Non-conventional Gas Resources

There are following non-conventional gas resources that have potential of varying levels for filling the demand and supply gap;

1. Shale gas
2. Tight gas
3. Stranded gas
4. Flared gas
5. Coal bed Methane (CBM)
6. Existing Conventional resources

There are environmental and climate implications to all these gas resources. CBM, if not utilized, can leak into the atmosphere. Methane has 35 times higher climate footprint than the most blamed CO2.Secondly, gas being climatically benign relatively, its higher availability in the energy mix would have impact on national climate footprint.

Comparative Gas Resources: Potential and Prices

	Resource(TCF)	Prices(USD/MMBtu)
Existing Conventional	20-25	4-5
LNG	Unlimited	6-8
Tight Gas	25-40	10-12
Flared Gas(150 MMCFD)	3	Free
Stranded Gas	500 MMCFD	0.25 USD over
Shale Gas	95-250	10-12
CBM	20	4-5

Sources: MPNR, OGDC, PPL and SCA

Shale Gas

It has been widely known that there are significant Shale gas resources in Pakistan. EIA reported Shale gas resources of 206 TCF out of which 51 TCF were recoverable economically. In 2011.More reliable estimations have been done recently in 2014 through USAID assistance which put Shale resources at 95 TCF of gas and 14 billion barrels .It has been estimated that Shale gas production cost would be 1round 10-12 USD per MMBtu, which under current market conditions is unaffordable, when LNG is costing around 6-8 USD per MMBtu. Other than cost, there are issues of heavy water consumption and its disposal, and its impact on agriculture and environment. Under current international and market environment, it is difficult to attract foreign companies. Technology is restricted in a few hands. MPNR has entrusted R&D projects to PPL and OGDC. Therefore, Shale gas remains at best a long term resource possibility. It appears that it is only through U.S. assistance that a major breakthrough may be possible in this area. One may have to wait for the right political circumstances to be able to attract major assistance from the U.S. in this respect. A US-CPEC initiative may be a goal worth pursuing for competing politicians and governments.

Tight Gas*

In simple words, it is a gas resource which is difficult to produce gas from. Gas is stuck in tight formations of low permeability/porosity and more drilling is required to extract gas resources. Consequently, higher prices would have to be paid to recover the production cost. Tight gas policy has been announced by MPNR which provides for a minimum of 40% premium over conventional gas prices which can go up to 50%.In order that such policy provisions are not abused(conventional resources may be pushed as Tight gas), an elaborated third-party determination process has been provided. Perhaps a review of existing policy and the elapsed time would enable softening down some of the unnecessary provisions. Half of the world gas resources are tied in unconventional resources such as Tight and Shale gas and CBM. Thus such resources cannot be abandoned. These resources would become increasingly important, once the existing conventional resources get close to exhaustion. Tight gas resources of 24-40 TCF have been indicated. Private companies have been demanding 10-12 USD per MMBtu which is twice the commercial LNG rates. MPNR has entrusted PPL and OGDC with the task of developing these resources based on which a realistic pricing formula could be developed. The recent drop in Oil and Gas prices have made it even more difficult to exploit such resources.

Stranded Gas Utilization

India has awarded recently a capacity of 5070 MW on stranded gas fields. E-Reverse Auction was used to determine the sales price of the electricity which came out to be IRs. 4.70(7.52 Usc) per unit.9 stranded gas fields have been auctioned with a gas capacity of 348 MMCFD.

In Pakistan, there have been attempts to utilize stranded gases and a Stranded Gas Policy-2013 has been announced, with the following features;

Reservoir size of Stranded Fields for Zones (Zones as marked in Petroleum Policy 2012):

 a. Zone I - [25] Bcf.
 b. Zone II - [20] Bcf.
 c. Zone III - [15] Bcf.

The other main issues are; 0.25 USD per MMBTU incentive; Government would be the buyer and third-party sales are allowed with provisions of sharing possible Wind Fall profits(over and above 8 USD per MMBTU);5 years implementation period has been provided. There is provision for virtual pipelines e.g., CNG transport. Reportedly, there are 14 stranded gas fields that are eligible under this policy with a total resource/reserve of 3 TCF. It is, overall, a good policy and despite its good features, nothing seems to have come out of it. The policy needs to be updated as since 2013, major changes have occurred in the national and international market, in particular, the Oil and Gas prices and induction of LNG in Pakistan. RLNG prices today are around 8 USD per MMBtu. The reasonable reference price should be the RLNG price for if you can import LNG at a certain price, why can't we award the same price to locally produced gas, especially the stranded assets. The main issue has always been as to who gets the business which usually prevents implementation and results in logjam. India has solved this problem as have other countries to invite open tenders. Competitive bidding also solves the problems of fixing prices and other parameters. So let us encourage market and competition.

---

*http://www.mpnr.gov.pk/mpnr/userfiles1/file/uploads/Tight_Gas_E_P_Policy_2011.pdf

*https://www.dawn.com/news/778516

Flared gas

Associated gas is produced along with Oil production in most cases. Sometimes, associated gas is not of quantity and quality to be processed and connected to the gas network. In such cases gas is either vented or preferably flared. Venting is very injurious to environment and methane being 35 times more potent to cause Climate effect than CO_2. Flaring is thus better than venting, although one would be better off with an economic utilization of gas than burning it off uselessly.

Globally, 3-5 % of the natural gas production is flared, which is a very large volume, enough to produce electricity for almost all of Africa except South Africa. In Pakistan, estimated flaring is to the tune of about 150 MMCFD, although no formal estimating has been done in this respect.

Flare gas policy has been announced in 2016 in Pakistan. Pakistan E&P policy, discourages gas flaring and allows it in exceptional, technical and economic circumstances. To quote, from the Gas Flare Policy, 2016:

"The Operator shall not flare Natural Gas but shall use it commercially or for recycling. If Associated Gas is not so used or not planned to be so used, the Working Interest Owners shall negotiate an arrangement making it available to THE PRESIDENT or its designee free of cost at the downstream flange of the gas/oil separation facilities in accordance with Article 11.4. If THE PRESIDENT for whatever reason is unable or unwilling to take delivery of the Natural Gas that would otherwise be flared as provided for above, the Operator will be allowed to flare such gas in accordance with the Rules without any royalty or excise duty liability until such time as THE PRESIDENT or his designee can take delivery."

Those investors and users who put the flared gas to good use are entitled to get the gas free of cost. In that case, it can be a good business if a viable use and technology is available. CNG appears to be a ready use and technology. Micro and Mini-LNG is not known as widely in this country. In both cases CNG or LNG, gas has to be processed, i.e., cleansed off the undesirable components to bring it to user or pipelines specs, as the case may be. Companies like Caterpillar and Shell have developed mobile systems which indicate the level of perfection and business potential that may be there. In Pakistan-gas is a small start-up which has started some activities in this respect. It is not known as to what level of success has been achieved by them. Recent drop in gas prices may have reduced their incentives. On the other hand, free raw gas availability should be a good enough an incentive.

Coal Bed Methane (CBM)

During coal-forming (coalification) process, a large quantity of Methane is produced, whereby the plant material is progressively converted to coal. This gas is trapped within the coal's internal surfaces and cleats. Coal stores 6-7 times more gas than the equivalent rock volume of natural gas. The evidence is the usual coal mine fire, which is actually generated by escaping Methane released during coal mining. CBM is not a mystery. It is being extracted many parts of the world. CBM is a clean Methane without much extra components and can be burnt directly in I.C. Engines and Boilers etc. Its CV is about 80% or more of chemical Methane.

In the U.S., CBM resource estimates have been put at about 700 TCF, out of which 11 TCF have been estimated in Lignite deposit of that country. Currently, 7 TCF of CBM is produced annually in the U.S. CBM is produced (embedded) at a rate of 100-700 SCF per ton of coal. More CBM is embedded and produced in hard coal and at deeper depths. Assuming a low value of potential release of 100 SCF per ton, Thar Lignite may have about 20-40 TCF of CBM. Preliminary studies indicate a potential output of 1-2 TCF per year as opposed to 1.5 TCF output of natural gas presently.

It should, however, be noted that CBM has to be produced before coal mining is done, otherwise CBM would be released into the air. And this is the usual practice. CBM wells are dug into the coal seams and gas is released by evacuating the trapped saline water. CBM thus can provide the energy needs of coal mining. Drilling for CBM is much easier and low technology process. In case of Thar, one may have to dig only 250 meters as opposed to 2-3000 Meters for Oil and Gas. Accordingly, CBM output is also limited; more than 10-20 CBM wells may have to be dug to get an output comparable to a typical natural gas well.

As has been mentioned elsewhere, Methane is 35 times more injurious to Climate Change than CO_2. CBM production can, therefore, earn CDM revenues as well besides producing energy. Exploration and development activities should be undertaken to accurately estimate the resource characteristics based on which a

CBM Policy could be developed. A small diameters plastic pipe network may have to be laid over a very large area to get a reasonable output. But certainly, it is worth the effort. The U.S. is extracting 7 TCF annually after all. An exploration contract for CBM in Thar had been signed many years back with a multinational company but was rescinded for unknown reasons. Renewed effort may be required to attract the right parties in this respect.

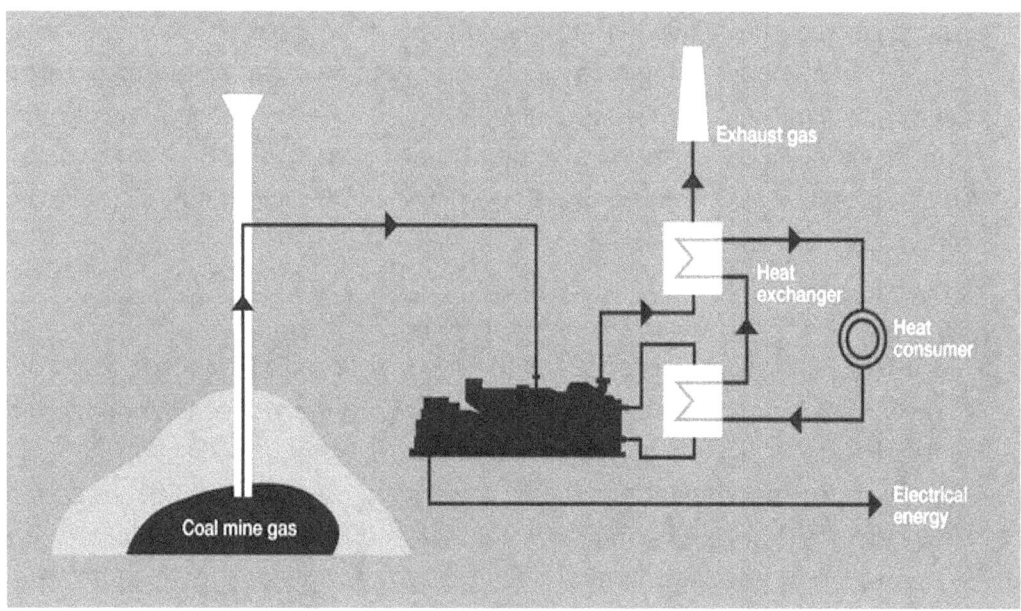

CBM burnt in I.C. Engines: Courtesy Clarke Energy

18. Alternative Approaches in RLNG

Five RLG plants have been planned. One has already been installed at Port Qasim. Contract for another has been awarded which is expected to come online within this year. Three more RLNG plants are in the pipeline, of which one at Gawadar is at an advanced stage of planning. Although there has been a meaningful progress and performance in RLNG, we would like to examine if there are alternative and better approaches in building further RLNG capacity.

Local gas potential is there, but immediate prospects of major finds and their development are not very promising. The dramatic reduction in International oil and gas prices has further enhanced pessimism. Law and order situation in Balochistan has not improved significantly enough to enable launching of a major exploration effort there. However, there have been success in KPK in discovering new gas fields and prospects for further developments appear to be attractive. But it takes about seven years to bring a discovered field's gas into the pipelines. With the advent of Donald Trump and recent aggressive statements against Iran, it appears that sanctions on Iran may not be lifted very soon, thus putting Iran –Pakistan Pipeline project at a distant future on the timeline. All of this indicates that the short to medium term future belongs to RLNG. In this timeframe, RLNG is expected to remain competitive.

In addition to long-term LNG supply agreement with Qatar, LNG is also being procured under spot arrangements based on international tenders. Controversy on this account appears to have been eased off, although it should be examined if more transparency can be brought in LNG contracts and the classical traditions of price secrecy even on the behest of credible suppliers like Qatar can be shunned. This has become even more important in the context of Panama leaks litigation. It is in the interest of all that the old commercial traditions are modified in keeping with the emerging local and international consensus towards accountability and transparency.

The two RLNG plants (Engro and PGPL[7]) at Port Qasim have been installed on lease basis. Engro has a tariff of 66 cents per MMBtu (rather too much) and PGPL has a Tariff of 41 cents per MMBtu, which is lower than a similar project in Bangladesh in which a tariff of 47 cents has been agreed to. The tariff given to Engro is excessively high, even though both projects are a result of international tender, which mandates some explanation. The most convenient explanation would be that experience and competition is bringing down the tariff. However, the fact is that there was much less competition in case of the second project (PGPL). There were two bidders, one was technically disqualified and the remaining single party got the deal. One would like to bring the future tariff rates of newer projects at more reasonable levels, in order to be able to bring down RLNG tariff at consumer end. It has been estimated that due to the existence of a number of intermediaries, the end tariff increases by 2 USD over the imported price of LNG (about 33 %).

Floating Storage Regasification Unit (FSRU)[8] has become very popular and almost a standard technology for RLNG projects these days. It takes lesser time, on the average, of almost one year to install FSRU, if the FSRU are in the market. Existing LNG transportation ships are also converted to FSRU rather swiftly. New FSRU, if built, would take more than two years. Leasing is always expensive as is readily discerned from everyday life experience. However, individual motivations and profitability of leasing are different than national and institutional ones. Leasing is definitely a very profitable business for lessors dealing with customers who may not have other choices and may have other benefits like tax deduction which justify even expensive leases.

FRSU of comparable capacities (650 MMCFD and 150,000 M^3 storage) these days are available at a net price of 270 million USD. Local expenditure charged shown by Engro of Jetty and pipeline investment is 125 million USD, (actual can be much less) taking the total to around 400 million USD. International lease rates of FSRU alone are around 130,000-150,000 USD per day, depending on leasing terms and the buyer and seller. However, total leasing charges (FSRU and local facilities together) per day are 272,000 USD in case of Engro and 255,000 USD for PGPL. It is take or pay contract, simply

---

[7] Pakistan GasPort Limited (PGPL), is establishing the country's second LNG import terminal. PGPL will provide storage and regas services to Pakistan LNG Terminals Limited (PLTL), a wholly owned subsidiary of Government Holdings (Private) Limited, for 15 years at a levellised service charge of $0.4177 per MMBtu, and 95-percent availability, on a take-or-pay basis.

[8] FSRUs take LNG and regasify it—taking it from the liquid form, where it is reduced in volume and expanding it back into a gaseous form where it's usable to make electricity for other uses.

meaning that irrespective of utilization level, these are fixed charges. The calculated tariff is notional, for comparative purposes only. If the utilization rate is less than the design capacity, the charges would be higher in proportion to the lack of utilization. It actually means, an annual fixed charge of around 100 million USD. Compared to an investment of around 400 million USD, a payback period of 4 years is a really good deal for suppliers.

My studies at Planning Commission have indicated that there is a margin of 100 %, which is devoured by RLNG developer and the lessor and there is a scope of reducing this surplus by either enhancing competition or introducing alternative financing such as debt financing. International loans are available at a rate of under 5% p.a. including all charges such as risk, servicing etc. In fact for shipbuilding, the lending rates are often subsidized by exporting country governments.

Alternative financing and even technology options have been explored and successfully implemented by India and Singapore and others. Lithuania, being a small and more conscientious country, has been trying to convert its lease agreement replaced by cheaper debt finance. However, it has not been successful, due to the unwillingness of the lessors. It is difficult to convert or change financing mode during mid-course, but it is certainly feasible in the beginning of the project, as is the case of our new projects. There are other cheaper options of buying old LNG ships (under 100 million USD) and having these converted into FRSUs.

If RLNG is to be a major cornerstone of our energy supplies as it is emerging, risks are to be evenly distributed. It may be a bit too risky to have all the RLNG projects on lease periods of around ten years. Some projects must also be on long-term ownership or IPP basis. On ownership-debt terms, one may have a tariff as low as 25 cents per MMBTU or even less, as opposed to 66 cents and 41 cents respectively for Engro and PGPL. The original Gawadar RLNG project had debt-financed IPP in mind. Apparently Chinese have withdrawn from it in the interest of locals who are seeing a lot of cream in it. In fact, leased based RLNG/FRSU is the easiest project to be installed, bringing in revenue and profit in one year only. It does not require much of an entrepreneurial know-how either. All you have to do is to find a FRSU lessor who handles everything except the local issues that might be there. Anybody having reasonable credentials is invited to benefit from my services in this respect. I am saying this in a lighter vein to emphasize my point, otherwise business deals are made by those who deal quietly in the dark of the night (as conspiracy theorists would put it) and not to those who write articles betraying which otherwise can be quite some commercial secrets.

Apart from alternative financing options, there may be more appropriate technical solutions. For example, Floating Storage Units (FSU) could be installed at a jetty and regasification facility can be on the shore. Old LNG ships costing under 100 million USD could be purchased outright to serve as FSU.LNG ships are always available not FSRUs. It may be more robust and less risky and even cheaper solution. A mix of approaches could be tried. The prospect of the FSRU sailing away leaving all our RLNG users high and dry appears to be rather scary.

There is a looming threat that future RLNG projects may be costlier making end tariff for RLNG even more expensive. We have already mentioned that there are a lot of intermediaries and overheads. For example, it has been proposed that a break-water (protection from wavy ocean) costing more than 100 million USD has to be made at Gawadar for locating the RLNG plant there. Although, I am not a port expert, it is surprising that despite all kinds of praise being heaped in favour of Gawadar location, a small place cannot be found on the coast that does not require expensive breakwater. A very small port has been built there which can hardly host a small number of ships. Where is the space gone? Reportedly, all good spaces have been planned for real estate interests. This is major issue obstructing industrialization in Pakistan. Industrial and commercial spaces are becoming unaffordable. How can we enhance our exports, if we keep making inputs expensive? May be the long-term spatial plans have to be more realistic and grandeur based on dreams than sober thinking. May be RLNG is being considered as risky in terms of safety as nuclear plants are. Most RLNG plants in the world have been located with a peripheral requirement of one km only.

Two extreme situations can emerge in case of infrastructure projects: one is of the low price , less than average incentives and an undersupplied market affecting economic growth as has been the situation in the period prior to 2013 and overflowing until now, and the other extreme would be of expensive projects under abundant incentives attracting developers and financiers resulting in oversupplied market and high services prices which are not affordable by the users, as has happened in Greece , Spain and elsewhere where governments have defaulted on debts and facilities e.g. railways go under- utilized as fares being unaffordably high .Both extremes are to be avoided and in an enthusiasm to end one extreme condition ,another extreme condition should not be permitted to be approached. Despite the purported sophistication, all lenders engage in careless lending. Often they are powerful enough both to afford losses and as well extract repayments.

Some leadership has to be shown by the people at the helm of affairs. There is abundant advice available, including the free one as has been offered in this space. It is not necessary to follow the oft-travelled path always. Enterprise and creativity has its rewards.

19. Alternative Approaches to LPG Air Mix Plants: Small Scale LNG Supply Chain

MPNR intends to install 65 LPG Air-Mix plants, at a cost of Rs. 17.6 Billion, to supply LPG-Air MIX to consumers in areas where LNG distribution network is not available. Under the scheme, LPG will be transported to such areas by LPG tankers already established in LPG supply chain in Pakistan. Liquid LPG will be gasified in LPG –Air-Mix plants and connected to the isolated distribution network installed for this purpose. LPG-Air-Mix would be supplied to consumers in designated areas at a tariff that is normally chargeable for Natural Gas (NG) supply. The relevant gas distribution company in its annual revenue requirements, to be approved by OGRA, would absorb the expenses.

Merit of the scheme is that far-off areas would be supplied with a viable fuel for normal household use, which otherwise would normally use wood-burning contributing to deforestation and both indoor and outdoor pollution. Quetta remained without NG for quite some time due to pipeline economics constraints, which contributed to national disharmony and political problems for national governments.

Criticism:

The scheme has been criticized among some circles on the following grounds:

- It may be quite expensive scheme both in CAPEX and OPEX terms and the commodity charges. LPG is as expensive as Petroleum products(11-12 USD per MMBtu).With additional expenses of LPG-Air-Mix Plants, it would become more expensive .Practically Gas companies would be buying LPG at 11-12 USD and sell at 3-4 USD per MMBtu. Won't it be better to sell RLNG of 6 USD and sell it at 3-4 USD per MMBtu? Even though such costs are being absorbed by other consumers through cross-subsidy tariff, it does cost in economic terms.

- Cross-subsidy may be opposed by OGRA.

- Legal issues may also emerge, as LPG and NG are two separate products and businesses.

Small Scale Liquefied Natural Gas (SSLNG)

Small Scale LNG Supply Chain (SSLNG) is an alternative solution. The main difference between LNG and SSLNG is that usually mainstream large LNG sector utilizes pipelines, while SSLNG utilizes Trucks and Trailers similar to LPG. However, containers/bowsers cylinders are designed differently in the two cases and are of different materials.

SSLNG is coming up again over the last few years after an eclipse period for a variety of reasons. It has a market share of 5 % in overall LNG market with an installed capacity of 20 MTPA worldwide. There are about 100 SSLNG production facilities and probably more than a thousand regasification satellite terminals. China, Japan, Spain, Portugal, Turkey and Norway are the leading countries in this field, while Latin American countries have also started taking interest. Remote areas, stranded gas fields and environmental reasons are the major drivers for SSLNG. Residential sector and transportation are important user sectors. Major challenge is high capital investment required in establishing complete supply chain, while for competing LPG sector most of it may already be there.

SSLNG range lies in between 50,000 to 1 Million TPA with a typical capacity of 400-1200 TPD. Ship transportation is limited to 30,000 M3.Both pressurized and chilled storages are used having farm sizes of 500-30,000 M^3. Pressurized Bullet Tanks may contain 50 M^3 LPG. BOG management or maximum pressure limitations define the user plant design approaches. LNG regasification may be more expensive than LPG Air-Mix plants. BOG may be produced at a rate of 0.5-1% per day. SSLNG is promoting LNG use in Transport sector whereby heavy truck and trailers are altered to use LNG.

The Alternative SSLNG approach

SSLNG has been working in several countries for the same reasons that have justified LPG-Air-Mix plants. LNG can be produced at existing NG fields that are geographically closer to the Supply areas. For example, Adhi and Neshpa fields could be ideal for this purpose. At Adhi, NGL (Natural Gas Liquid) extraction plants are operating already. Small scale LNG plants can be integrated with NGL plants ,that already have the Cryogenic facilities and other common infrastructure creating scale economies and saving energy consumption .Such LNG production at NGL sites is already being done. Standard processes have been developed by reputed companies like Linde, Flour and Bechtel etc. in this respect.

LNG thus produced can be transported thru LNG tankers/containers to the relevant areas where small scale RLNG (regasification) plants can be installed in place of LPG-Air-Mix plant. The other steps of distribution would remain the same.

SSLNG can be coupled with mainstream LNG in the South and South West of Pakistan, whereby LNG from PQA or Gawadar LNG terminals may be off-loaded to LNG trailer in liquid form and transported to remote areas where pipeline network may not be there and urban-rural settlements having small populations to ever justify pipelines. A Milkman approach may be used supplying/offloading smaller amounts to mini-RLNG systems.

For Northern areas, gas fields in the forward areas like Adhi and Neshpa may be utilized for LNG production and shipping. Stranded fields may also be evaluated for the same.

Advantages of the proposed System

Admittedly, the proposed system may not be applicable to all areas but would be limited to the areas where Gas fields are available nearby .LNG can, however, be produced by installing small scale Liquefaction facilities with or without NGL facilities. NGL makes it cheaper only. LNG can be potentially cheaper than LPG. If some stranded fields are available, the attractiveness of the proposed scheme is even enhanced.

LPG is normally priced equal to Gasoline, unless differential taxation may create large differences between the two prices, as is usually the case in Europe and to some extent in Pakistan. RLNG through SSLNG operations may be cheaper than LPG but more expensive than mainstream LNG prices due to scale economies. Where wellhead NG (gas) prices are low as in Pakistan and probably in China, RLNG of SSLNG may be cheaper and competitive with LPG. For exact determination, detailed cost studies are to be made.

Table 26 Small Scale LNG(SSLNG) perspective data and parameters

SSLNG Definition (range)	MTPA	0.05-1
Market Size	MTPA	20
Market Share	%	5
SSLNG Ship Transport	M3	30,000
SSLNG Tank Farm	M3	500-30000
SSLNG Liquefaction capacity	tons per day	400-1200
SSLNG Regasification Plants	tons per day	400-1200
Boil off Gas(BOG) production	% per day	0.5-1
NG Well head prices	USD per MMBtu	3.01
LNG prices landed	USD per MMBtu	6.01
LNG prices users end	USD per MMBtu	9.01
LPG Prices PRL	USD per MMBtu	10.79
LPG Prices off BYCO	USD per MMBtu	11.91
Household Tariff NG	USD per MMBtu	0.74-3.71
Average Tariff SNGPL	USD per MMBtu	3.41

Source: OGRA and others, compiled by the author

Integration of LSLNG and SSLNG

Figure 8 depicts the proposed integration of LSLNG and SSLNG. Direct LNG off-take can be taken from RLNG terminals being built at PQA and Gawadar. While LSLNG would send RLNG into the country's main pipeline system, under SSLNG LNG off-take would be taken from the RLNG terminals directly from LNG ships without resorting to or passing through Regasification. LNG would be loaded onto LNG Trailers and transported to far-off locations where isolated distribution network has been installed along with a small Regasification facility. Users are supplied RLNG from this distribution network.

Table 27 Comparative LPG and HSD Prices (Whole Sale/Ex-Refinery)

	Price	CV	Price
	Rs per ton	MMBtu per Ton)	Rs per MMBtu
LPG Price Byco	48906	45.326	1078.98
LPG Price PRL	54000	45.326	1191.37
LPG Price Cylinder	93500	45.326	2062.83
HSD Price ex-Refinery	56251	44.045	1277.12
HSD conversion factors		1135	49.56
Average NG Price WH			3000

Source: OGRA

Apart from small distribution networks, LNG can be supplied to LNG/CNG stations in the mainland. Trucks and Trailers running on LNG/CNG may get filled from these stations. Apart from environmental advantages vis-à-vis HSD, there is normally a cost saving of about 30%. Long-haul trucks running on LNG may require refills at a distance of 500 kms and thus many and frequent LNG stations are not required. China, Turkey, EU and the U.S. have established LNG truck fleets in large number and have been reportedly working satisfactorily.

Ship-to- small ship transfer of LNG can also be provided into the RLNG terminals, whereby small ships get filled from the larger ones. Small ships may supply LNG to small regasification terminals installed along the coastline and serving far-off areas.

Small-scale Gas liquefaction facilities can also be installed on Gas fields closer to the far-off regions and LNG supplied which after transportation may be regasified and distributed. Stranded Gas fields may be utilized in this context. Specifically speaking, such systems can be installed at gas fields like Adhi and Neshpa etc.

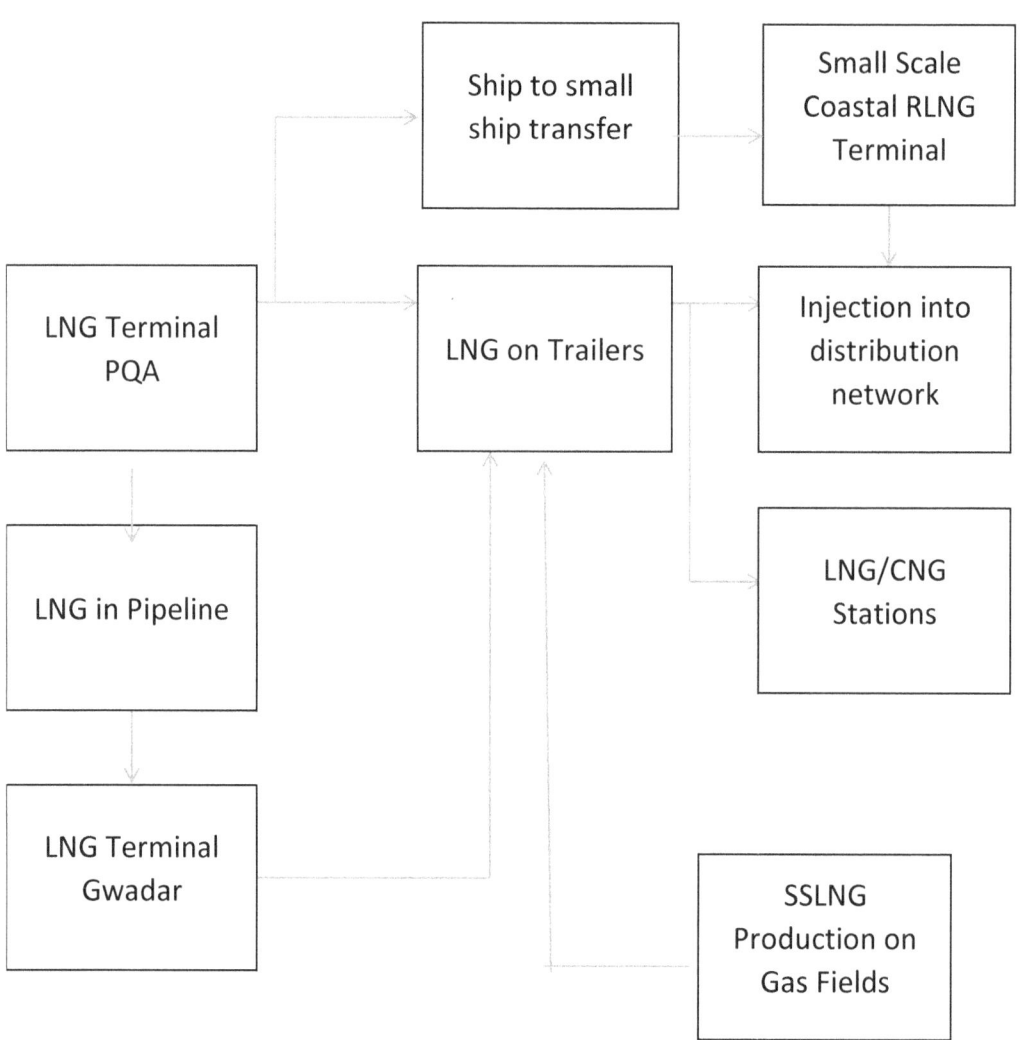

SSLNG vs. LPG Air-Mix

As mentioned earlier, LPG can be as expensive as HSD costing around RS 2000 per MMBtu at retail level. Under the proposed LPG-Air-Mix system, Gas companies would be buying LPG at around Rs.1200 per MMBtu and after adding 600 Rs per MMBtu in transportation and other costs, would be selling a commodity (LPG) which may have cost the GASCOMs around Rs 2000 per MMBtu, at a normal NG gas tariff of Rs 77 to Rs 350 per MMBtu. Under the proposed SSLNG, GASCOMs would be buying LNG from RLNG terminals(PQA and Gwadar or from Gas fields) at a price of around 600 Rs per MMBtu and after adding transportation costs, it may add up to Rs 800 per MMBtu which is still a lot lower than Rs 2000 per MMBtu cost of LPG-Air-Mix. Small scale Liquefaction on Gas fields may also cost around Rs 800 per MMBtu.

LNG/CNG Stations

Direct LNG off-take from RLNG Terminals (PQA, Gwadar) can be transported in LNG Trailers to LNG/CNG stations. Existing CNG stations can be retrofitted where space is available and large user trucks can be accommodated. LNG in these stations can be stored to be sold and filled in LNG vehicles or alternatively it can be converted into CNG through heat exchanger and without utilizing Mechanical Compressors thus saving energy.

LNG production at NGL Plants on the Gas Fields

Pakistan's large gas fields can earn additional revenue besides solving Natural Gas logistics problem with respect to far-off areas by producing LNG, which then can be transported to these areas by LNG Trucks and Trailers. We have indicated elsewhere that LNG imported (or even Local through the proposed process) is going to be far cheaper than LPG which is normally sold at Gasoline/Diesel price equivalent.

Cryogenic NGL extraction plants from natural gas offer an attractive opportunity for LNG production as a side stream. Cold gas is tapped from the overhead of the de-methanizer and liquefied via a compact and efficient mixed refrigerant compander process. The enthalpy balance of the NGL plant is maintained by recycling a cold liquid to the NGL column.

The standard concept extracts about 200 tons per day of LNG from the residue gas of a NGL plant with a capacity of 200 MMSCFD. Very favorable economics can be demonstrated, as in addition to the LNG production the NGL plant capacity is boosted by approximately 10% (AIChe).

All those fields, which have installed NGL recovery (from Gas stream) Plants, can benefit from it. For example, Adhi, not among the largest gas fields with a gas output of 61MMCFD can produce about 50 tons of LNG per day. Other gas fields, which have much larger output (Qadirpur 426 MMCFD, Kunnar-Pasaki 114) and run by OGDC can ,in principle, produce proportionally larger output, in case they extract NGL under Cryogenic process.

It should be clarified here that almost all the gas output can be converted to LNG, in principle, but here we are concerned with a portion of gas output that can be converted to LNG in conjunction with NGL plants resulting in much lesser energy consumption of about 0.5 kWh per Kg of LNG output. Secondly, it is not additional LNG output, but the equivalent gas would be diverted from the main gas feed. There is a general lack of awareness resulting in a perception that LNG production is highly energy intensive, which however, consumes only 8-10% of the feed gas.

Conclusion and Recommendations

1. Admittedly LPG-Air-Mix may be a simpler system that is based on the existing LPG transportation assets, however, it is certainly very expensive in terms of commodity charge, which is equal to or even more than Gasoline or HSD.

2. LPG –Air-Mix may be better suited to Northern areas wherein close by LPG sources may be utilized.

3. SSLNG may be better suited for South and Southwestern Pakistan due to close by LNG terminals of PQA and Gwadar.

4. Existing CNG stations can be retrofitted with LNG systems. Compressors would be redundant, as Heat Exchanger would create CNG pressure. The retrofitted ones can serve both CNG and LNG vehicles.

5. All RLNG Stations should be obliged to provide for SSLNG infrastructure to be able to provide up to 10% of RLNG capacity in the form of SSLNG (LNG unloading Arm and Trucking facilities etc); additional cost to be charged at cost-plus formula.

Figure 13 LCNG Stations

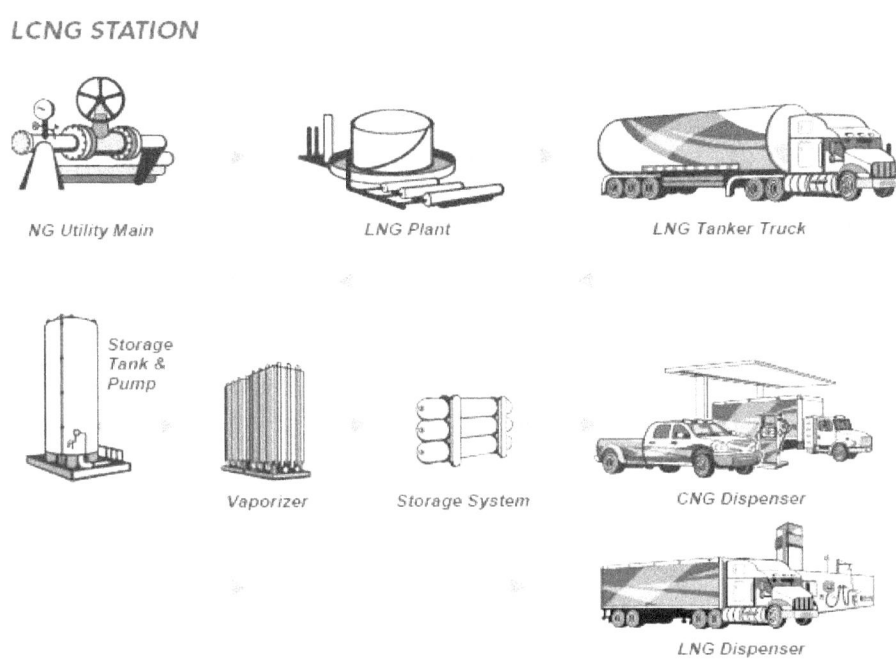

A liquefied-compressed natural gas (LCNG) station combines LNG and CNG in one station. A typical LCNG station is supplied with LNG and has dispensers for both LNG and CNG vehicles. Like an LNG refueling station, an LCNG station relies on a local LNG supply that can be delivered by tanker truck, similar to diesel and gasoline. The advantage of an LCNG station is that it can offer both LNG and CNG. This type of station can also be set up in areas where there is no local natural gas distribution.

At an LCNG station, LNG vehicles are fueled in the same way as at an LNG station with a cryogenic pump moving the LNG from an insulated storage vessel through a dispenser into the vehicle. To produce CNG, the LNG is pumped into a vaporizer that converts it from liquid to gas in a controlled way so that it can be dispensed at the right pressure as CNG.

20. LPG subsidy Scheme for averting deforestation

MPNR has launched a scheme of installing off-grid distribution networks for distributing LPG. Some 24 schemes are being planned. The idea and intention is good and is to be appreciated, however, the approach and methodology does not appear to be attractive and viable. In this space we will examine the issues involved in the proposed scheme and would explore other options as to their feasibility. In an earlier writing, I examined an alternative approach of distributing Truck-transported LNG to such off-grid networks, which is used in many countries. In this space, I will explore another option i.e. that of LPG cylinder scheme under a subsidy scheme.

Only 20% of households in Pakistan have access to a clean and convenient cooking fuel like piped natural gas. Another 1.5 million Households have access to LPG cylinders, which leaves most of the country unsupplied, except for the major urban areas. Thus the idea of supplying some kind of clean fuel in affordable prices should be received with sympathy and affection.

There are two types of subsidies involved in the proposed scheme. One that a LPG of 9 USD per MMBtu (FOB Saudi Arabia) and 13 USD per MMBtu at site would be supplied to consumers at a tariff of about 2.8 USD per MMBtu, the latter being the most widely applied tariff for small domestic users which are in the majority. The deficit would be procured through cross-subsidies meaning thereby other category of consumers would be charged higher to cover this cost. Industry may not like it, but they should understand that it would ultimately benefit industry if the workforce is able to cook its food without polluting his home and endangering his and his family's health. Industries do not grow in a vacuum on the back of half fed populace. Those industries and societies which competitiveness are built and based on inhuman living conditions do not last or prosper very long and perish as we are seeing the condition of our textile industry, producing cheap and low quality low-priced goods which is matched by still lower prices offered by more inhuman countries called competing and competitive. One would rather not have such industries. I apologise for some digression, but probably well deserved, I understand.

Coming back to the main issue, LPG is more efficiently distributed in cylinders worldwide. I do not know of a country where such schemes of LPG distribution through pipes is practiced. There are apartment complexes and gated-societies where such contrivance is used. But it is rare that LPG is distributed through piped network to whole towns having several thousand plus customers. The costs should be presumably higher and more safety risks are involved. One usually finds leaking gas outside and inside houses. But it is natural gas which being lighter goes up and is diluted to safe levels in the atmosphere. LPG has heavier molecules and tends to reside and flow close to the ground after it has had opportunity to leak. And thus is more susceptible to combustion and explosion. One keeps hearing LPG origin fires and explosions more than that of natural gas. The proposed scheme tends to cater to the elite who in their big houses would like to have piped gas and not the cylinders. The poor may still be outside the periphery in which LPG distribution network would be installed. Poor is spread out much more widely. And look at Balochistan, where population dispersion is so wide that it may not be even possible to take electric wires, not to talk of LPG pipe which is even more difficult.

However, I am even a bigger supporter of supplying LPG to as many people and localities and to the poor at a subsidised rate. India, our arch-enemy, spends 2 USD per capita on LPG subsidies. In 2014-5, total LPG subsidies amounted to 2.365 million USD. For example's sake, if we decide to adopt the same model, we would be spending 400 Million USD on LPG subsidies for the poor. In India, one cylinder per month per household is supplied under subsidy schemes; the subsidy depending on the poverty levels could vary from 50 to 100 %, meaning thereby that some eligible users may get one LPG cylinder per month literally free.

I would propose a LPG subsidy scheme of 100 million USD per year to be partly financed out of budget and partly out of cross-subsidy with the Natural gas sector. Government companies such as PSO and PPL may import LPG under this scheme which would enable them to import something like 250,000 tons of LPG in a year (684 tonnes per day) at a prevailing rate of around 400 USD per tonne. This would mean about 21 million cylinders per year or 1.75 million cylinders per month. By comparison, it has been estimated that there is a current demand of 90,000 tons per month in summers and 115000 tons in winters, out of which 40% goes to domestic consumers.

PPL and PSO would get it free under the 100 million USD per year subsidy and would charge the filling and transportation and handling charge. The eligible poor households will have to get registered with an exclusive distributor appointed by PSO and PPL, possibly to be centrally computerised utilising NADRA system. There is always a possibility of misuse and abuse as has happened in India. Some misuse is to be tolerated. However, there are other ways of passing on this targeted subsidy, of directly transferring a monthly amount of subsidy to the accounts of registered users, but on the placement of order to the local distributor, so that the targeted subsidy is utilised for the purpose ,it is meant for. The system is reporting well in India. We can adopt variants that are suited to our conditions. It is expected that one would be able to supply one LPG cylinder at a subsidised price of 300-400 Rs. Provincial and federal governments may be required to share subsidies under some formula. Current unsubsidised price is around Rs 1000 per Cylinder (0.531 MMBtu or 11.8 kg LPG per cylinder, 18.83 USD per MMBtu).

Initially, the target should be the northern areas including rural parts of Balochistan, KPK, FATA, AJK. Giligit-Baltistan and Chitral etc. LPG for heating and cooking is a necessity in those parts of the country, although the cold is so much in those regions that no amount of subsidised LPG may be enough. However, cooking needs can be covered under the scheme by providing one cylinder per month. The public benefit of this subsidy in addition to the improvement in living conditions of people would be lesser tree-cutting avoiding or reducing deforestation.

LPG supplies under the proposed subsidy scheme will reduce deforestation which has been fast eroding forest cover in the country. Current deforestation rate is 1.63-2.1 % per year which is the highest deforestation rate in Asia. As per erstwhile MDG , Pakistan was required to increase forest cover from 2.5% to 6% by the year 2015.It may be possible to attract international funding towards the proposed LPG subsidy scheme focussed on reducing deforestation(REDD).It would be relevant here to quote from a kfW study on the subject.

In studies of villages in the Himalayan region of India carried out among communities where LPG use had been encouraged by government action, fuelwood use decreased from 475 kg per capita per year in 1980-85 to 46 kg per capita per year in 2000-05, suggesting that LPG can play a very significant role in forest preservation in a low-income environment (110). The energy savings from total fuelwood requirements for cooking in these villages were estimated to be as high as 70% in lower level Himalayan altitudes (3,742 MJ per capita per year), but considerably less in the higher altitude regions with considerably less LPG usage and greater heating needs (111). In Senegal, the growth in LPG use in the 1970s resulted in the avoided consumption of about 70,000 tonnes of fuelwood and 90,000 tonnes of charcoal annually (equivalent to 700,000 m3 of wood per year). The Ministry of Energy estimated a 15% decrease in deforestation rates due to LPG adoption.

Source: Source:https://www.kfw-entwicklungsbank.de/PDF/Download-Center/Materialien/2017_Nr.7_CleanCooking_Lang.pdf

Concluding, it may be worth re-examining the existing LPG scheme piped networks under implementation by the MPNR and substitute it by this proposed scheme of subsidised LPG distribution in cylinders.

Table 28: Proposed LPG Subsidy scheme-Major data and parameters

Monthly LPG consumption	
Summers	90000 tons
Winters	115,000 tons
Annual LPG consumption	1-1.2 million tons
	101 million cylinders per year
LPG import Price	400 USD per tonne or Rs 40 per kg
	472 Rs per Cylinder
Local LPG Retail rate	Rs. 90 per kg or Rs. 1062 per Cylinder
Proposed LPG subsidy	100 Million USD per year
LPG bought under Scheme	250,000 tons per year
Cylinders sold under subsidy	21.186 million Cylinders of 11.8 kg
Market Share of Subsidy LPG	20%
Retail price of Subsidy LPG	300-400 Rs per Cylinder
No of LPG Cylinders per year	12 cyld per household
No of Household benefitted	1.76 million Households
Total HH in Baloch and NA	5 million HH
Fuel wood saved	2 tons per year per Household
Fuel Wood/Trees saved	3.5 million tonnes per year

21. LNG Controversy

LNG contract with Qatar has become a hot political topic. The contract was handled by the P.M. Abbasi when he was Minister of Petroleum although he continues to hold the charge of Ministry of Energy which has recently been formed to merge Ministry of Petroleum and Ministry of Power. The antagonists of the PML (n) and PM Abbasi argue that Pakistan and Qatar LNG contract prices are unduly high and there is a possible scope for corruption and rentier intermediary interest in the contract. We would like to present and examine the facts and data to let the readers make their own judgement about the merits of these allegations. I must clarify at the outset that there is no intention here to support or oppose any partisan view.

Pakistan LNG contract is a government to Government contract for the supply of LNG for a period of 15 years with an annual average supply of LNG of 3.75 million tons, valued at 1.3 billion USD per year at current prices. The contract was negotiated after a considerable negotiation process. Qatar has been a dominant and up to a certain time a kind of a lone supplier for many countries. Competition in LNG sector has emerged over the recent years. American Shale gas has come into the export market and Russian Gazprom have come into the market. LNG prices have come down along with the drastic reduction in oil prices. Before Oil price reduction LNG prices were around 14-16 USD per MMBtu which now have come to 5-7 USD per MMBtu. Unlike oil, LNG prices are still regional and there is no standard and uniform reference price for LNG unlike Brent crude oil index prices. Lowest LNG prices are in the U.S. based on cheap Shale gas and in Europe LNG prices are also lower due to cheap exports from the U.S. and competition of LNG with pipeline gas.

Pakistan Qatar contract was finalised sometimes in 2015 which was in earlier part of the emanating competition. The contract price formula was agreed at 13.37 percent (Coefficient technically) of the Brent Crude oil price which translates to around 6.68 USD per MMBtu, assuming current typical oil price of 50 USD per barrel. For those who may not be initiated in Oil and gas pricing, let me clarify here that LNG prices are not 13.37 percent of Oil prices. In fact these are around 75-80% of oil prices. The figure of 13.37 is a multiplier for convenience sake. When one multiplies the crude oil price (which is in USD per barrel) with the coefficient of say 13.37, one gets the resultant LNG prices in USD per MMBtu.

After entering into contract with Qatar, MPNR invited tenders as a result of which ENI an Italian company offered the minimum bid prices of 12.26 % which were further negotiated down to 11.9% of Brent Crude. There is a considerable difference between ENI prices and those of Qatar of the order of about 11%.ENI could offer this kind of low price through combining many sources cheap and not so cheap. Qatar prices are higher comparatively with other company prices because, firstly, these are negotiated prices and are not tender prices, secondly, Qatar thinks that it offers a more secure and lesser-risked contracts in terms of the security of supply.

However, Pakistan-Qatar LNG prices, if compared with the contract prices of Qatar with India and Bangladesh, appear to be almost the same or even lower. Let us examine the data as provided in the adjoining table 27.Although the coefficient-multipliers of India and Bangladesh are lower than Pakistan –Qatar contract, there is an additive number in the contracts of the two countries. This number when added makes their effective prices higher than that of Pakistan. Thus Indian and Bangladesh prices which are perhaps a more appropriate and relevant reference than Japan or Europe.

We have mentioned earlier that LNG prices are regional and vary among the region and also depend on the market power and volumes of orders. Japan's low prices of 5.8 USD per MMBtu can be seen in that perspective. European prices are even lower at 5.28 USD .And that of the U.S. still lower. Thus it would be in appropriate to compare LNG prices in international terms as we do in case of oil.

Additionally, SPOT prices are not comparable with 15 years long term contract. SPOT prices vary under supply and demand and also vary seasonally and may be affected by temporary or occasional events. Under SPOT prices, supplier has almost no price risks; he supplies at prevailing prices. Also availability of LNG is not certain under Spot prices. This is true for LNG than for crude oil of which supplies are usually abundant. If RLNG based power plants are to be fed, one cannot rely on spot availability; what one may possibly gain in terms of possible lower prices, one may lose under mandatory capacity payments under irregular spot supplies. Under long term contract, although there is a link with Crude Oil prices, there may be price variations beyond that are covered under the formula. Longer the term of the contract higher the price risk and higher the prices. We have another example of Guvnor prices offered to Pakistan for a period of 5 years which were still lower than ENI prices, which proves our point. It has been decided by GOP that there would be no more government contracts for procurement of LNG and that private sector would be allowed to do that.

The real winner is Pakistan ENI contract at a price of 5.95 USD quite closer to that of Japan. Ironically and interestingly, both ENI and Qatar contracts have been done by the same agency and headed by the same person-Minister-PM Abbasi. The difference is that one is negotiated and the other is the result of comparative bid. I have always argued in favour of competitive bidding as I have been doing in case of Solar and Wind Power which prices have been coming down under competitive bidding. However, as explained earlier LNG is slightly more complicated.

According to an article published in April 2016 issue of OGFC (Oil & Gas Finance Journal), some one-third of LNG capacity is booked by portfolio companies such as ENI, Shell, and Total etc. Some of this may be sold under long or medium terms contracts to utilities or other buyer companies, and some of it may be available for Spot sales. I am not privy to, if MPNR had some expert advice available at the time of negotiating LNG contract with Qatar, it appears safer to deal with them or benefit from their advice, as we have seen in the case of lower price contract with ENI. Seller countries themselves always have rosy appreciations of their products and may not be as flexible and market savvy as these portfolio companies are.

Ideally, one would have liked to have the same prices as that under ENI contract or Japanese METI spot prices. t would, however, be unfair to compare a later day ENI contract with that of Pakistan-Qatar. LNG sector and prices are evolving like that of Solar, although in less dramatic way. There are people who compare higher tariff of QA solar park with today's lower solar PV prices of 6 cents or lower. The whole world bought expensive Solar PV at higher prices earlier. It is always a difficult question when to enter and that what the future prices would be. While it may be arguable as to whether the timing was right for QA Solar,

LNG decision could not have been postponed due to the energy crisis. One might prefer Gas-on-Gas contracts such as e.g., average of the last three months Japanese METI Spot prices instead of Oil linkage. We are a small buyer and Japanese are still buying at oil linked contracts.

Table 29: Comparative LNG Import Prices in Countries

	Brent Coefficient	Constt	LNG price * USD/MMBtu	Effective Coeff.
Pakistan -Qatar Gov to Gov	13.37	0	6.685	13.37
Pakistan ENI	11.9	0	5.95	11.9
India-Petronet-RasGas-Qatar	12.66	0.6	6.93	13.86
India-Petronet-Exxon-Australia	13.9	0	6.95	13.9
Bangladesh-Qatar	12.65	0.5	6.825	13.65
Japan METI-2017 SPOT Avg yr			6.4	

*based on 50 USD Brent crude oil price
Source:1)DNA India;2)Daily Star Bangladesh;3)METI-Japan;4)MPNR-Pakistan

Finally, why should there be a secrecy on contract. I am no privy to any confidential decisions. However, the relevant people would be foolish enough to bring in such questionable confidentiality, if there is some hanky-panky. In that case what is the plausible reason? The contract prices are declared now and communicated both to OGRA and PSO. Qatar government might have insisted on keeping the contract or/and price secret so that it does not complicates the price negotiations with other parties. We have seen that Pakistan prices are slightly lower than offered to India and Bangladesh. I am not sure how advisable it was for Pakistan in these circumstances to go with the secrecy provisions. In any case, it is no secret any more.

Concluding, Pakistan-Qatar contract prices are comparable with those in comparable and relevant countries in South Asia. LNG prices are a complicated issue and not as simple and standard as Crude oil prices. There is no standard and uniform reference price as is there in Brent Crude accepted widely. LNG market is evolving and prices may come down. However, we would be bound by our long term agreements. Long term agreements have their pros and cons as everything else has. If price differences increase, there may always be a possibility of renegotiations within or outside the contract at political level. The foregoing still does not prove that somebody may or may not have made money in LNG deal. Third party facilitators having long term service or sales contract may always be able to earn some money. That money, however, comes out of the sellers pockets. Sellers normally do not give a benefit of non-involvement of a facilitator to the buyer.

RLNG Local Selling Prices

Table 28 provides data on local selling prices of RLNG. Reader would note that a considerable overhead is added onto the basic import price of LNG. Large customers are given delivery off the transmission pipe and thus carry much lower price mainly on account of lesser losses.

Table 30: RLNG Prices Sept 2017(USD per MMBtu)

	Transmission	Distribution
LNG DES charges	6.9761	6.971
Service charges	1.4762	1.4813
Losses (%)	0.9200	9.8900
Losses	0.0700	0.8358
Total without GST	8.5224	9.2881

Source :compiled by the author based on OGRA data

National Natural Gas Market Overview: World LNG Landed Prices

Federal Energy Regulatory Commission • Market Oversight • www.ferc.gov/oversight

World LNG Estimated Landed Prices: Jul-17

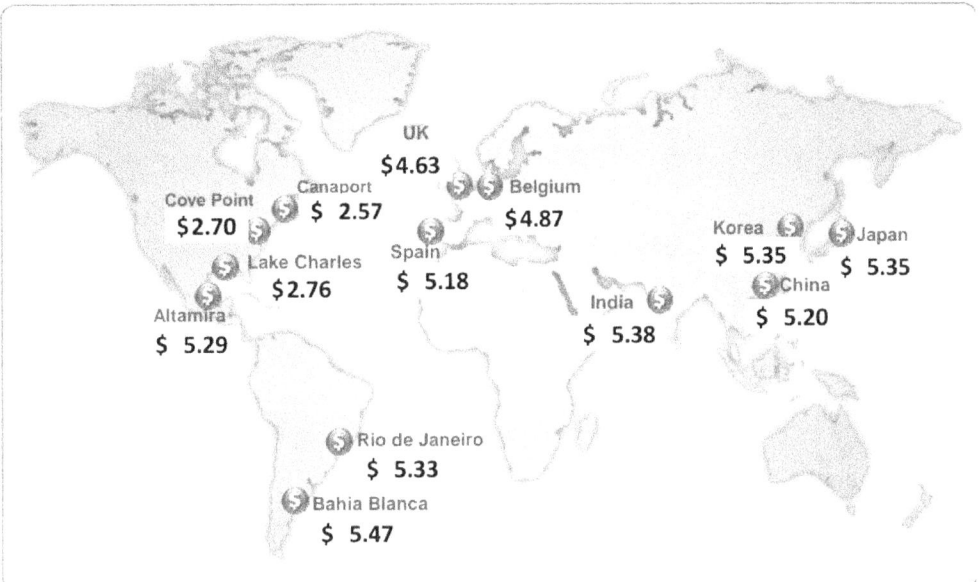

Source: *Waterborne Energy, Inc.* Data in $US/MMBtu
Note: Includes information and Data supplied by IHS Global Inc. and its affiliates ("IHS"). Copyright (publication year) all rights reserved. Prices are the monthly average of the weekly landed prices for the listed month. Landed prices are based on a netback calculation.

Updated: Aug-17

22. Curtailing Losses in Gas Sector

T&D losses in the gas sector have assumed alarming dimensions amounting to 15% of the gas handled by the two companies, namely; SSGC and SNGPL. The monetary value of these losses have been estimated at around 8-900 million USD calculated at an opportunity cost rate of 6 USD per MMBtu of LNG prevailing price. These losses can be reduced and savings to the tune of 500 million USD can be generated, if corrective measures are taken. Therefore, it is worth the precious time of the readers to go through the submissions made in the following on how to go about achieving such loss reduction.

Considering the importance of the important yet contentious issue being frequent subject of court litigation, OGRA recently commissioned a study by engaging consultants who have come out with extensive recommendations for dealing with the problem. These recommendations are these days subject of heated discussion within the gas industry and will form to be a substantial part of this writing as well.

Compared to power sector where losses are around 20% now, gas sector losses used to hover around 7% which over the recent years have more than doubled to around 15%.OGRA has been allowing about 7-8 of these losses to the companies resulting in considerable loss of revenues and thus the litigation by the companies as mentioned earlier. In the jargon of gas industry, such losses are called UFG-unaccounted for gas. We will be using these terms interchangeably in our discussion. Apart from monetary losses, these losses should be of extra concern as our local gas resources are dwindling. There are mainly two components of losses in both power and gas sector which are; 1.Technical 2.operational.Technical losses are the ones that may be reduced but not avoided altogether. In power sector, it is what is called Copper loss caused by the resistance of wires and cables. There is no equivalent to this in gas sector; second part of technical losses are measuring errors which in gas sector is more complicated.

Regulators have almost a thankless job of treading midway between the consumer and producers interest while ending up annoying or dissatisfying both. Economic efficiency argument can be a saviour of the regulator and in fact for all, if that criterion is used in addition to the usually political decisions. It is not an easy computation, therefore, one starts with comparing the loss(UFG).In most developed economies UFG are around 2-

3%,while in Bangladesh, Turkey and astonishingly Russia and Texas, UFGs are around 5%, a figure around which consensus seems to be emerging to be the reference in our case as well.

Certainly, the current level of UFG reflects a deplorable state of the gas industry, for which it alone is responsible. However, rejecting them without any help or assistance to come out of it would be counterproductive. High UFG results in higher tariff which hurts domestic consumers' pockets in a poor country and affects competitiveness in export markets. Pakistan's textile exports are allegedly suffering due to low gas prices prevailing in Bangladesh, although party would be over in the latter country as its gas resources are also dwindling and has started importing LNG at a price more than twice that of locally produced gas.

Consultants have grouped the losses (UFG) into the following seven categories;

1. Normal Theft (by consumers and non-consumers)
2. Theft in poor law and order conditions and areas(in tribal areas and KPK, certain areas simply break the pipelines and supply gas to themselves making political claims on gas resources in and around areas where gas has been found and is being produced. Even gas fields have been attacked and taken over by local political representatives and notables; one may have to put up with it till a suitable political solution is developed .Hence ECC allowed such losses to the companies).
3. Minimum Billing issues
4. Measurement Errors (including billing inefficiencies which consultants seem to have overlooked, but in this scribes view, could be quite a significant factor)
5. Leakages(due to older pipes more than 20 years old, lack of full coverage by cathodic protection systems and interrupted operations of such systems due to frequent power failures, etc.)
6. Sales Mix Issues(ratio of bulk sales to domestic sales has decreased causing more losses than earlier due to larger fragmented control difficulties, although it should affect cost of service more than the losses)
7. Others

In my view, the major issue is the large geographical territories being controlled from headquarters in Karachi and Lahore. Consensus has been emerging that these two companies be divided into smaller companies' ala DISCOs in power sector. In fact, there is a case for dividing MEPCO and PESCO having very large geographical territories under their command. The issue has been complicated by additional proposals of separating transmission and distribution functions. The issue is further complicated by the nature of ownership; although the companies are majority government owned, these are listed in stock exchanges, making companies assets divisions even more difficult. Foreign consultants and IFIs meddling also complicates and even impedes implementation, otherwise, the Minister Abbasi was of the opinion of implementing most of the urgent reforms and restructuring in six months, otherwise, as he believed the windows of opportunity do not extend over longer period of time. How prophetic he was as we see the unravelling political circumstances.

Table 31 Actual UFG vs. Allowed UFG in SSGC and SNGPL over the recent years

SSGC	2011	2012	2013	2014	2015
Gas available for sale(BCF)	396	406	418	423	434
Actual UFG %	9.4	10.8	10	15.4	15.2
UFG allowed by OGRA %	7	7.5	8.6	8.6	8.6
SNGPL					
Gas available for sale(BCF)	665	675	638	582	522
Actual UFG %	12.5	11.5	13.4	13	13.5
UFG allowed by OGRA %	8.3	8.1	6.7	6.9	7.1

Source: OGRA-KPMG study

A common problem in both power and gas sector are inadequate control metering. Meters are to be there at all nodal points in order to be able to do control accounting. It is not enough to know the losses at company level; there has to be control metering for every small geographical segment ,say, of 1000 customers enabling the respective managers to know how much gas has been shipped and how much billing and receipts are there in their respective areas. The difference can help identifying and zeroing on problem areas. Higher management can also order controls, investigations and campaigns based on such information. Incentive schemes can also be devised based on such data and information. Readers may be surprised to know that only 12% of 4058 TBS nodes are metered in SNGPL and 26%(out of 2442) in SSGC. Besides, SNGPL also seem to lack in the number of TBS points as a percentage of the number of consumers. There may be a case for smart metering of these nodes as well, as we have argued elsewhere, in case of power sector, to start with installing smart meters on Distribution transformers, in order to be able to implement a smaller and affordable programme in a shorter period of time. However, much improvement and control of theft is possible by simply covering all TBS points with meters.

Replacing leaking pipes and old meters and installing meters at all control points such as TBS requires money. Cash starved companies cannot invest money and nobody lends to such failing companies. Hence, it may be advisable to provide a reasonable tariff and UFG allowance. However, it is a double edged sword. A liberal UFG may provide further incentives, opportunities and market for company employees to collude with gas thieves. Hence whatever be the UFG, it is to be accompanied by a programme of loss reduction and action plan that identifies activities that have to be done. And that is what consultants seem to have done in there study. They have recommended a two part UFG allowance; 5% fixed and 4.05% variable under an action plan spreading over a period of 5 years. In the year 2015, SSGC had been awarded a UFG of 8.6 % and SNGPL of 7.1%.The proposed action plan consists of 25 actions, 15 of which are technical requiring investments and 5 are related with operational activities required for direct theft detection and control.

However, a few aspects are lacking in the consultant's study(although, they are restricted by the TOR and the resources provided), in my view, and can still be undertaken; firstly, the investment requirements of the 15 proposed actions to enable the companies plan and to enable OGRA to provide some tariff incentive and assistance to the companies, if required; secondly, comparative economics of gas transmission and distribution activity is a must to be able to see the deficiencies in a perspective. Russians and other investors have been demanding a tariff of USD 0.75 to 1 per 1000 cft for transmission pipelines alone. It would be useful to know and compare the T&D charges of gas companies with other companies in the world, making UFG allowance and concessions decisions a bit more objective and meaningful. Thirdly, I wonder if it would have been possible to estimate the losses in each of the seven categories of losses identified, may be, three broad categories. This would have enabled target setting in terms of output as well as opposed to input based targets proposed by the consultants. Both targets have their own advantages and disadvantages. Fourthly, it may be examined if unit rate tariff setting would be more advisable than the annual revenue requirements approach, the latter being the current practice. And as a special variant of the unit rate, a constant price tariff ala KESC power tariff may be workable in a similar gas situation wherein loss reduction is to be incentivised. KESC constant price tariff (with provision for escalation and adjustments) developed by ADB consultants, has worked well encouraging the utility to make investments for improving its thermal and operational efficiency. KESC could also come out from red into a profitable venture. Fifthly, Gas is 35 times more injurious to climate change than ordinary CO_2 which is a usual subject of discussion and mitigation. Leakage reduction programme could earn a continuous stream of revenues under CDM (Clean Development Mechanism) for the next 20 years, if leakage estimates and investments plan thereof could be worked out.

23. Promoting Oil & Gas Production

PM Abbasi has been doing well in all other Energy sectors except local production of Oil and Gas. LNG infrastructure and supplies have been developed and more are in pipeline, while new oil and gas pipelines are being installed. This has already made an impact on energy supplies. It is said that there is no gas load-shedding for industries these days, and gas supply to domestic sector would be better this winter due to commissioning of yet another LNG terminal; one working already and another in offing.

Foreign oil companies never had much interest in Pakistan, especially, after 1970s.Most of the oil and gas fields are owned by local public sector companies like OGDCL and PPL. This indicates that they do not expect major oil discoveries compared to the Middle East in Pakistan. However, most studies indicate Oil and Gas potential that may be sufficient to cater to the needs of Pakistan. More of Gas has been indicated than oil. More than 150TCF of gas potential has been indicated as against a demand of 8 BCFD (Billion Cubic Feet per day), which may be enough for another 40-50 years.

While imported LNG is doing well for fulfilling immediate needs and solving the energy crisis, almost a continuous and permanent reliance on it, will create problems of trade deficit and may be foreign debt as well. Take or pay liabilities can also be treated as debt or near debt, as it will result into payments abroad irrespective of level of production. There is almost exclusive focus on LNG, so much so that organisations are being floated in LNG sector unnecessarily. While Pakistan LNG Ltd should have been enough for handling the supply chain of LNG ala PSO dealing with Oil, another company with the name of Pakistan LNG Terminals Ltd (PLTL) has been floated, causing nothing but confusion and expenditure. This reflects state of mind, a mind fixated on LNG exclusively.

It would be unfair to say that nothing has been done to promote local oil and gas production. Some effort, more as inertia and momentum of the past, has been done. Over the past three years, 319 wells have been drilled with 91 new finds. As compared to the past, this is a good record and performance, perhaps the highest than ever before. But it is not enough compared to the demand challenge and supply potential that exists. It does not match with the progress in the power sector; 20,000 MW Installed total up till 2013 and 10,000 MW in the period after. An initiative much larger in scope and strategy is required, as we shall see later in our proposals.

OGDCL and PPL have been full of money all these years. They are making profit even now under USD 50 per barrel, not to talk of the profit made in the days of over 100 USD per barrel oil prices and consequent gas prices, although gas prices are lower in Pakistan due to price ceiling, but there is no such ceiling on Oil. They have been so much flush with money that a separate company had to be made to park the funds of OGDCL. Oil and Gas sector has money collected under various head such as Gas development Surcharge (GDS) as well and other pockets of money. Interestingly and amusingly, due to this money flush factor, MPNR floated a proposal in 2014 to go into power production projects as well, a move that was amusingly rejected by all government institutions.

The solution lies in creating market and competition and resolving governance issues. Look at what wonders have been done and are being done in KPK by KPK Oil Company. It has come out of autonomy of action and creating a third actor, apart from selecting competent persons to man the organisation. Half of the 21 active drilling rigs are reportedly working in KPK by KPKOGCL, although done by PPL and OGDC. KPKOGCL is targeting 2 Billion CFT per day of gas production by 2025. Let there be more entities. For example, MPNR should actively consider creating one or two more companies ala PPL and OGDC. An Oil drilling contractor company and a JV in oil service area. Local Private sector investment by smaller companies may be promoted after all they earn and lose in stock market Sattha daily. In Pakistan, one strikes one discovery for every three wells, not a big risk. Unfortunately, service companies are leaving Pakistan, as is the recent example of Baker Huges indicates. It may be due to overall restructuring and also because country presence is not considered necessary by the service companies in the information age. It costs 15-20 million USD to drill and complete an oil or gas well in Pakistan, which is 3-4 times more than elsewhere on the average. It can be brought down by creating a market of supply chain in Oil construction and service industry. Well construction costs would come down by creating the infrastructure and the market. Rigs can be partly produced locally except the drilling bit and rotary equipment. Local supply of rigs and installation know-how can be promoted. By these measures, existing companies can double their drilling output and possibly double the discoveries.

Finally, MPNR has to loosen its clutch on the sector. There has been a demand by the provinces of having their own petroleum concession departments which in their view would have promoted efficiency and output and reduce the time lapses in decision making. Demand has been met by only federalising the governance of the petroleum concession department inducting representatives from the provinces. In my view, no harm would have been done by provincializing the petroleum concession department. Balochistan may have the capacity issues which MPNR can provide by acting as a facilitator. MPNR can keep policy making unto itself and pass on implementation i.e. actual award of concession to the provinces. After all, what is the fuss about; much ado about nothing. It is about awarding concession to explore and develop. All policies are in place. King Saud IbnulAziz is said to have signed concession agreement in a day and now they are enjoying (post script: It is learnt that decision has been made recently to devolve petroleum concession to the provinces).

24. Enforcing Environmental Fuel Standards to get rid of Smog

The whole of Punjab is under smog these days paralysing life and business therein. It has been a routine affair for this time of the year lately. NASA satellites have revealed that pollution from crop stubble burning has crossed into Pakistan exacerbating the situation. Although this is not the subject here, the episode indicates the need for cooperation and consultations in the region and thus of the right political climate to enable the aforementioned. The discovery of the influx of pollution from India need not make us oblivious of our own environmental responsibilities, as influx is only a part of the problem. In the following, we will examine what we can do through stablishing and enforcing environmental fuel standards in the country, for bulk of the pollution and thus the smog is generated through burning one kind of fuel or the other.

Before coming to the fuel standards, let us first take up the issue of crop stubble burning which is done on both sides of the border. Farmers burn crop stubble to clear the residual crop matter in order to be able to seed the new crop. This not only creates smog but also contributes to the climate change. Thus there is dual incentive to take care of this. The cheapest way to clear the excess crop matter is to burn it. However, machinery can be employed to cut and collect the biomass for other uses; alternatively, crop matter can be recycled by cutting and pulverising the biomass and let it be mixed in the soil.

Farmers are perennially short of money .They have to be provided credits and training to utilise appropriate equipment .This biomass can also be collected and burnt adequately in boiler furnaces of the smaller power plants to be installed nearby. Many studies have been done in this respect and require implementation.

Let us come back to the fuel standards. In Pakistan, high Sulfur fuels have been used both in power plants and in automotive. The pollution problems are exacerbating due to influx of new vehicles on the road at a fast pace. Fortunately, Furnace Oil power plants are being closed down due to the influx of environmentally benign LNG. In the case of motor fuels, such as gasoline and diesel, not much care has been given to the sulphur control in these fuels.

The whole world is shifting to low sulphur fuels. In Pakistan, only lately, we have been able to partly switch to low sulphur fuel satisfying the requirements of Euro2 standard, which is a rather old standard limiting sulphur level to 500 ppm. Most of the world has shifted to or in the process of shifting to even lower sulphur as required per Euro5 or 6 limiting sulphur to only 10 ppm. Industrialised countries have a problem of converting their refineries into low sulphur mode. However, we import most of our fuel from abroad and should not face problem in such switching. Low sulphur fuel is available in international market. It is a bit expensive but compared to the health and environmental costs, such extras are much less. India has already adopted Euro4 (sulphur level 50 ppm) and is poised to switch to Euro5 or 6 by the year 2020.Thus our gasoline and diesel will be having 10 times more Sulfur and thus 10 times more sulphur emitting vehicles on our roads, even after the complete switch over to Euro2 standards. I would argue for adopting 50 ppm standard (Euro4), and not be victims or hostage of the existing old refineries which cannot produce low sulphur fuels. Either they should be asked to import low sulphur crude oil or their output should be relegated to smaller towns and rural areas. All major urban areas should be required to be switched to 50 ppm sulphur standard through distributing imported low sulphur fuels.

Secondly, in many parts of the world, ethanol mixed gasoline is being marketed.E10 gasoline contains 10% ethanol and E5 contains 5%.In the U.S., almost all cars have switched to E10,mixing mostly bioethanol made out of sugar cane. This has an added advantage of having a RON rating of 94 or 95.As reported in the press recently, manganese and other metals have been added to raise the RON levels. Manganese harms both the engines and the environment. We are exporters of Bioethanol produced as a by-product of our sugar industry. Many years ago, PSO did introduce E10, but more as a benign nonstandard fuel which people did not buy and PSO discontinues, may be for other reasons as well. However, E10 is no more a novel fuel but is a standard fuel .Our relevant organisations such as OCAC,OGRA,EPAs under the tutelage of the Fuel division of Ministry of Energy should sit together and examine both the issue of E10 and as well as of low sulphur fuel.

Coal power plants have been unfairly blamed for the current smog problem. Although many coal power plants are being installed in the country, currently, only one plant has been commissioned and that too in Sahiwal. That plant too has been working at a plant factor of 50% as per latest NEPRA reports. Thus, it is a misconception. However, the potential of coal power plants to cause harm is great, if low sulphur coal (1% S) is not used and appropriate environmental equipment such as SOx, NOx and Particulate matter controls are not installed. Fortunately, NEPRA has mandated 1% coal and the employment of controls. There is a provision of online monitoring as well. Responsible agencies should see to it that all such controls are actually implemented, as there is

Table 32 Environmental Motor Fuel Standards in India

Date	Particulars
1995	Cetane number: 45; Sulfur: 1%
1996	Sulfur: 0.5% (Delhi + selected cities)
1998	Sulfur: 0.25% (Delhi)
1999	Sulfur: 0.05% (Delhi, limited supply)
2000	Cetane number: 48; Sulfur: 0.25% (Nationwide)
2001	Sulfur: 0.05% (Delhi + selected cities)
2005	Sulfur: 350 ppm (Euro 3; selected areas)
2010	Sulfur: 350 ppm (Euro 3; nationwide)
2010	Sulfur: 50 ppm (Euro 4; selected areas)

Source: Ministry of Fuel, India

always a tendency to reduce costs in the business sectors. We should avoid extremism; one group opposing coal altogether and the other soft=pedalling the environmental controls issue. Thus it is possible to keep adding vehicles and power plants, but so long as the environmental care is adopted, we can save ourselves from not only smog but can afford to have healthy air in our neighbourhoods. Civil society organisations and the government would have a major role in getting the environmental provisions implemented.

TablE 29 provides the evolution of Sulfur control standards in India.They have opted for the ultimate Euro 6 standard for the year 2020,limiting Sulfur level in the motor fuels to 10 ppm only which would be comparable with anywhere in the world.

PART IV: COAL

25. The Problems and Prospects of Coal?

Sahiwal coal power plant (1,300 MW) has been commissioned and is working satisfactorily .My salute to Chinese that they have built a power plant of this capacity in such a short time. Government of Punjab has also been involved in solving administrative and logistical problems and deserves appreciation. Another similar coal power plant is at an advanced stage of construction and should be commissioned in the next few months. In Thar, Engro's Coal power plant is slowly coming up as well.

It is probably a stage where we should undertake some critical examination of some of the issues that have come up as decision about new projects is to be made. These projects were done in haste and it was natural that some due processes were avoided and lesser time was available to deliberate. I will try in this space to raise these issues and make some submissions and recommendations, although some of these I have been raising even earlier. These are the issues of cost of generation and CAPEX, environmental concerns and role of EPAs, NEPRA proceedings, imported vs local coal, location of future plants. Let us see if I can do justice with all in the following.

Adequacy of CAPEX

CAPEX and cost of generation of these coal power plants has been quite a burning issue. It has been argued that it is high and that, elsewhere, in India it is around 5 cents as opposed to 8.36 cents in Pakistan and that CAPEX should have been 1 million USD per MW instead of **1.45** MnUSD/MW for these plants. It is a very complicated issue. It is not like a soap whose costs and prices can be compared, even in that there is a lot of difficulty these days due to the variety that is available. There is a very large cost variation among various projects in the world; of timing, locations, fuel cost, logistics, interest rates etc. Fortunately, we have data compiled by a respectable international institution (IEEFA), which can be compared with our projects that has been reproduced in the adjoining table. It should be noted that the CAPEX figures are of some very reputed market players like Doosan, Alstom and Mitsubishi and should be comparable with CAPEX of China and vice versa. And these CAPEX figures include a full set of environmental controls like ESP (for particulate matter), FGD (for Sox control) and SCR (for NOx control). Locally developed and built plants in India are much cheaper i.e. less than 1 Million USD per MW. These are without environmental controls except the installation of ESPs. On the other hand, as it appears, the danger is that the same might be true for Chinese supplied projects in Pakistan. If that comes out to be the case, it would be highly unfortunate.

My judgment is that for plants having complete environmental controls, our CAPEX of 1.45 Mn USD /MW is rather excessive, and there are add-ons like Jetty costs and inland coal transportation. Had there been competition (of which there was no possibility under the financing arrangements under CPEC), some improvement was possible in this respect. However, the issue is that are environment controls as promised have been provided. It is a major controversy today, which we will take a little later.

A typical and median CAPEX figure is one million USD per MW, as we have provided data from IEEFA study elsewhere. In China itself, the prices are even lower.

PowerMag (http://www.powermag.com/who-has-the-worlds-most-efficient-coal-power-plant-fleet/)reports a recent project of 2x1000 MW complete with stringent environmental control built at an installed cost of 0.6-0.8 million USD/MW in 22 months. Greenpeace reports similar CAPEX figures, which table we have provided in the adjoining. Admittedly, Green Peace report provides overnight costs. Even if IDC and other costs are added, the total would be more than 0.8 million USD per MW. There is a strong case for bringing down coal power tariff in new contracts. NEPRA would be well advised to undertake a fresh determination diligently which is due any way. We have raised the issue of unduly high ROEs and additional miscellaneous costs elsewhere, which along with moderation of CAPEX can bring down coal power tariff of new plants to a fair and reasonable level. The process can be handled pleasantly without annoying the investors. If facts and data is presented logically, it is very difficult for parties to reject it unilaterally.

Improving NEPRA Norms

However, some improvements (reduction in coal power tariff) is still possible. A 16% IRR can easily be brought to the standard 15%. Reportedly, Chinese would have been happy to get 15%, as against a standard of 12% in this industry. Secondly, there are a lot of miscellaneous costs such as Insurance and financing fee where due diligence is required by NEPRA. A 3.5 % of financing fee amounting to 42 Million USD under a government-to-

Table 33 Capital Costs for Construction of New Coal Fired Power Plants (US$mn)

Project	Country	Technology	Technology Supplier	Completion	Capacity (GW)	Cost (US$ bn)	Cost (US$ mn/GW)
Manjung	Malaysia	USC	Alstom and CMC (China)	2015	1.0	1.2	1.2
Tanjung Bin-4	Malaysia	SC	Alstom (France), Mudajaya and Shin Eversendai (Malaysia)	2016	1.0	1.1	1.1
Mae Moh Power Plant	Thailand	USC	Alstom (France) and Marubeni Corp (Japan)	2018	0.6	1.1	1.8
Kudgi STPP	India	USC	Doosan (South Korea) and Toshiba (Japan)	2016-17	2.4	2.3	0.9
Khargone TPP	India	USC	BHEL and Alstom - Bharatforge (France, India)	2019	1.3	1.5	1.1
Bellary TPP	India	SC	Alstom (France) and Siemens (Germany)	2016	0.7	0.7	1.0
Yermarus TPP	India	SC	BHEL (India), Alstom (France) and Siemens (Germany)	2016	1.6	1.7	1.1
Gadarwara TPP	India	SC	BHEL (India)	2017-18	1.6	1.7	1.1

Source: IEEFA, CEA India

Table 34: China CAPEX Estimates Coal Power Plants of various capacities

Coal Type	Plant Type	Capacity(MW)	CAPEX(MCNY/MW)
Gangue	District CHP	2x350	4.58
Gangue	Electric Power	2x660	4.33
Gangue	Electric Power	2x1000	3.84

Overnight CAPEX, exclude IDC or any financing Cost
Is China doubling down on its coal power bubble? - Greenpeace
www.greenpeace.org/.../
Greenpeace_Doubling%20Down%20on%20Coal%20Power...
by L Myllyvirta

Government financing arrangement is not understandable. This kind of financing fee is charged when there are scores of financial sources lending in packages of 50 million USD etc. and many agreements have to be drafted and negotiated. It should do its job more carefully. Thirdly, Chinese financial insurance (Sinosure) of 7% of debt/project cost (another 84 million USD) should have been absorbed in the spread margin of 4.5 % that is already there for this purpose. Where is the good-will of CPEC? All of this could have been negotiated down had some sensitivity been shown.

Also we will have to leave this Hatam-Tai type of generosity giving somebody, 15% and others 16%, 17% and even 20% IRR based returns. Market rates are 10-12 % in energy sector. 15% should cover all the extra risks etc. We have dealt with this in more detail elsewhere. I haven't seen any formal study on this subject. All decisions have been arbitrary and simplistic. To be fair, NEPRA has occasionally shown some toughness and sensitivity. For example, in case of HVDC-4000 MW, NEPRA has shown courage and stuck to their guns.

Thar coal Transportation to Karachi

Under this laissez-fare, our local investors are proposing all kinds of impossible projects: Thar coal being transported to Karachi/PQA and power plant being built there. Transportation is cost-plus. You can propose your plant anywhere and you will get reimbursed generously. The argument being given that, Thar has capacity for 10,000 MW of power plants due to water and other reasons. At first, the bonafides of this statement have yet to be verified as there are many options and scenarios. But amazingly, how far-sighted have we become that we are thinking of such a remote future when 10,000 MW would be installed. And such projects are being pushed for priority. My point is that if this kind of light-mindedness continues, an already questioned tariff would become even more objectionable. I have seldom seen any lignite power plant away from the mine-mouth because of its low calorific value. However, if at all Lignite is to be transported, it should be towards north, to border areas of Punjab like Taunsa and others. In the instant case, there is no sense of bringing (energy) coal to Karachi and then make heavy investment in transmission of electricity. Another curse of Inland Freight Equalization Charge (IFEC) is being borne..

Why Thar Lignite is twice as expensive compared to elsewhere?

Another issue of importance is the imported fuel and the future foreign exchange drain. Uncertain future of the prices is unknown. The only way-out is local production of gas and coal and a good mix of renewable energy including hydro, solar and wind in order of emphasis. Gas will be required even if all power is shifted to non-fossil for domestic and industrial uses. Local gas is half cheaper than imported LNG and will remain so. Returning to local Thar coal, as to why it is twice more expensive than similar lignite elsewhere? Firstly, it's costing is based on cash-flow; secondly, a very high return of 20% IRR has been given, quite contrary to international and even national practice; thirdly, more than one coal mines are being started almost together at roughly the same time. I am not sure who will pay for the upfront cash-flow. Better would have been to permit one mine let it acquire its maximum output and then second mining project should have been initiated There is still opportunity to do this and decision makers ,provincial ones who own the resource and federal ones who deal with electricity)should sit together and bring some order into it. Fourthly, adequate due diligence on the part of regulators could have brought down the quoted costs. Fifthly, antiquated mining technologies are being used. Nobody in these days tries to excavate a depth of 220-250 meters with shovel and trucks. Bucket Wheel Excavators (BWE) are used, especially, in Lignite mining. India is using BWE for the last 50 years. Partly, Engro's precarious financial situation and induction of new partners made them start with Shovel and Truck so show performance. At 20-25 USD per ton, Thar coal would be a good deal, comparable with cost elsewhere and it would save foreign exchange and promote employment. Some steps can still be taken for cost reduction and corrections be made for the new mining projects. First of all, we don't need new projects that badly, secondly, some control advice or coordination should be made with Sindh Coal people to cut down their generosity on the shoulders of the consumers who would be paying ultimately.

The risks of imported Coal

Half or more than half of the generation tariff is the fuel cost. One coal power plant of 1,300 MW would result in a coal import of around 300-350 million USD depending on the going price of fuel. If Government of Punjab's heresy is accepted and many more coal power plants are installed in Punjab, then the coal import bill alone will be 2.0 billion USD along with an equal or even more of debt-servicing. Three RLNG plants are already being installed to meet their persuasion and another RLNG plant has been approved by the Federal Committee on Energy only a few days back (today it is 10th June 2017).Imported coal, apart being a foreign exchange drain, would cause unnecessary load on the Railways and the roads and will spread coal dust pollution all around the passage.

Coal is contrary to the expectation is getting expensive as a ratio of oil price. There is international opposition to call asking for all kind of taxation. Even, India has doubled taxation on coal to finance solar, may be. South Korea has levied a tax of 22 USD per ton on coal. In Australia, Victorian government has tripled royalty on Brown coal. There will be increasing opposition and taxation on coal making it expensive, if not being totally out of market. On the other hand, Solar and wind are coming under or around 5 cents. In nutshell, no imported coal projects should be allowed anymore. Federal ministers should put their foot down in rejecting demand of imported coal projects that keep emanating from government of Punjab. Federal decisions should be made in Islamabad only. It would heard national solidarity if this kind of practice is continued for long.

Another Imported Coal Power Plant announced in Rahim Yar Khan

Government of Punjab, after inaugurating iwal coal project, announced yet another coal power plant based on imported coal. Many people in this country have raised eyebrows on it, to say the least. In my opinion it is a bad decision, which will hurt them and hurt us all. Let me tell you, why?

Interestingly enough, RYK is a better site than Sahiwal and perhaps, if at all, coal power plant should have been located at RYK first in the first place, as RYK is closer to Karachi and it would have been far less costly in financial and environmental and railway congestion terms. Also perhaps, Rahim Yar Khan is less densely populated and agriculturally productive than RYK, although I am not quite sure about it (a separate research is required on this aspect, until then we rely on intuition and general perception).

Another issue is that are we going to forget our Thar coal resources which most of the people have been craving for. Two coal power plants, one in Sahiwal and the other in Karachi, each of 1320 MW, should have been enough to meet the emergency. Further coal plants should have been planned and installed in Thar. It costs 4 US cents of foreign exchange to produce one unit of electricity on imported coal and thus one coal plant would be loading an annual import bill of 400 million USD.

The RYK coal plant in Punjab and elsewhere should be CFBC based power plants ala Engro's Thar coal power plants instead of Pulverized (PC) coal power plants that have been installed for imported coal. This would enable them to use local Thar coal as well or a mixture of local and imported coal. The damage would thus be minimized. RYK, being on Sindh border, is close to Thar than any other site that is being considered in Punjab. Also, CFBC boilers may not need, although I am not absolutely sure, FGDs that are being avoided so meticulously and ardently by the combine of IPPs and the energy establishment. CFBC boilers have come of age and are available in large capacities these days and have acquired the required thermal efficiencies comparable to those that have been installed herein.

One could contemplate a direct road or rail link between Thar and RYK and then onwards to Sahiwal. From Islamkot in Thar to RYK, there is a distance of 700 km by car and 500 km crow flight (point to point straight) distance avoiding the curvature. A special road link can be built or preferably a rail track to connect Thar (Islamkot) with RYK coal power plant. The same route can be used for taking the Thar coal to Sahiwal, in case Sahiwal coal power plant can be converted to Thar coal or a mix of imported and local. In some hard days when there would be foreign exchange problem, this solution can be handy. But solutions take time, as we have seen, it is taking almost 5 years to solve the load-shedding issue.

Punjab government is pursuing energy autonomy objectives almost like a separate country without realizing its implications on foreign exchange and others and Sindh government continues to behave as a resource monopolists, little realizing that times have changed. They should in these circumstances act as resource marketers offering better terms and prices. Instead they keep insisting on Higher Returns such as IRR of 20%, which reduces Thar coal competitiveness. Also they are awarding one mining license after other without regard to market conditions. They should have waited till existing Engro mine is booked up to the optimum required capacity resulting in cost reduction.

Concluding, RYK coal power plant is a big mistake. Apparently, this is a victory of government of Punjab vis-à-vis federal institutions especially the minister and the MOWP who has been advising and insisting on the Punjab government to use Thar coal and avoid further plant on imported coal. But this victory would be an ultimate defeat as it would hurt them and hurt us all eventually. Would they desist from it, reading the afore-mentioned? I am not sure.

What to do with GENCOs: Coal, Lignite or RLNG?

Another related issue is what to do with GENCOs, when new capacity is available by 2018.GENCOS are highly inefficient, consuming expensive Furnace Oil or Gas. There are several proposals; the oldest one is to convert these to coal; another is to install RLNG plants; third is to close these down except the newer ones and convert these to either gas or coal. A case by case analysis has to be done. It would be better to simply shut down the old and inefficient. And conversion to imported coal, in my view, would be a great mistake. The merit of converting to Thar coal may, however, be examined in one or two cases.

The Prospects of Clean Coal

It is clear that coal is being abandoned, generally, and in Europe and the U.S. in particular. No new coal power plant is being built in these regions. In fact, none has been built in this century. However, there are limited prospects of HELE (High Efficiency, Low Emissions) based coal power plants. In South East Asia, including Japan and Korea, there is still some momentum for coal. Japan has not shun coal. Reportedly, it has 45 coal power plants in active planning and schedule for next ten years. However, these would all be on HELE with a guaranteed efficiency of 45%, as opposed to 39% average of Supercritical technology, which has been implemented in Pakistan. India will not be able to totally shun coal power and similarly the developing countries of Thailand, Malaysia, Indonesia, Philippines and Vietnam.

HELE is becoming more of a norm than a mere good thing. India and China have toughened both efficiency requirements and emission standards. We have provided a comparative list of new and old emission standards for developing and advanced countries elsewhere (Table 30, Chapter 25).Malaysia has recently (March 2016) commissioned a coal power plant (Tanjung Bin-T4) which has an efficiency of 47% under EPC contract with GE Power. In Poland, Jowarzno III (1345 MW coal power plant) has been installed with an efficiency of 45.91%. In Germany, Karlsruhe Coal Power Plant (912 MW) has a net thermal efficiency of 47.5%.In Denmark, Vattenfall has installed a 400 MW coal power plant (Nordjylad) with an LHV efficiency of 47%.Isogu in Japan has an efficiency of 45%.It appears that the target of 50% would be achieved very soon by the industry in order to survive in a hostile anti-coal environment.

Table 35 Summary of NEPRA Tariff: PQEPP Coal Power Plant at Port Qasim/Sahiwal

		660 MW	2x660=1320 MW
Capacity	MW	660	1320
Net Capacity	MW	607	1214
Auxiliary Consumption	%	8	8
Net Thermal Efficiency	%	39	39
Capacity Factor	%	85	85
CAPEX(total)	Mn USD	956.1	1912.2
CAPEX per MW	Mn USD	1.449	1.449
o/w Sino-sure Insurance	MnUSD	63.9	127.8
Financing fee	Mn.USD	21	42
Exchange Rate	Rs per USD	97.1	97.1
Interest Rate Foreign	%	LIBOR+4.5%	LIBOR+4.5%
Interest Rate local	%	KIBOR+3.5%	KIBOR+3.5%
Debt Equity Ratio		75/25	75/25
Debt	Million USD	717	1434
Equity	Million USD	239	478
IRR	%	16	16
RoE	%	27.2	27.2
Annual debt servicing avg	Million USD	91.34	182.68
Annual Equity Profits	Million USD	65.008	130.016
Imported Coal landed cost	USD per Ton	129.6	129.6
Annual Coal consumption	Million Tons/yr.	2.055	4.11
o/w Coal FOB	USD per Ton	90	90
Reight	USD per Ton	20	20
Capacity Charge(CPP)	UScents per kWh	3.6448	3.6448
Energy Charge(EPP)	UScents per kWh	4.7153	4.7153
Total Tariff(TPP)	UScents per kWh	8.3601	8.3601
Source: NEPRA			

Source: NEPRA Tariff Determination

We have emphasized in the foregoing the need of assuring that all environmental standards be enforced and the required equipment be installed, as the CAPEX and Tariff have a provision for it under a very high price of USD 1.45 per MW. In the new procurements, all the lacunae and confusion should be removed in this respect. Also, an efficiency standard of 45% should be adopted. These are the issues and not mud-slinging. Coal power is new to Pakistan, expertise and knowledge is lacking and those who have the expertise are kept at an arms distance under the system that we have. Things had to be done on fast-track. But, there is time now to make corrections and compensations. If the suggested actions are taken, GoP would get more appreciation and praise by the people and by the history.

Coal Jetty Costs: Are these too high?

Coal jetties are required to unload imported coal from ships. Coal jetties / terminals are required even in surface transportation of coal. There are three issues, (one we have already discussed above) on which there still is confusion or controversy. One still does not know (in public domain) as to what would be the impact of railway transport of coal on tariff. In a public hearing, Pakistan Railways maintained that NEPRA did not have jurisdiction on Railway Transport tariff. Apparently, the issue has not been resolved yet, as NEPRA has kept quiet on it. On the other hand, Pakistan Railways has not revealed its transportation rates.

We will focus here on Jetty costs, both capital and operating costs. NEPRA has awarded a tariff component, on the account of jetty costs, of 9.86 USD per ton based on a CAPEX assumption that is quite contentious, as we shall demonstrate. NEPRA has also allocated 10% of FOB Coal costs as other costs, which are presumably the operating cost of the coal terminal and may be incidentals. This also works out to be another 9 USD per tonne (based on the FOB coal costs assumed in NEPRA tariff determination). This adds up to a good 18.86 USD per tonne, more than the freight cost of bringing coal from Indonesia or South Africa. This, in my opinion is excessive, as I shall demonstrate in the following.

NEPRA has been confused on this issue from the beginning and the stakeholders could not guide NEPRA on this as well. There is a limit on as to what stakeholders can provide to NEPRA. First there is a vested interest (among those who know) not to tell the real data. Secondly, coal power is new to Pakistan and there is a lack of knowledgeable people on the subject. Free data and info has its limitation. Expert third party consultants have to be engaged for true assessment of costs. Competitive quotations can be a charade as well. CAPEX estimates in cost-plus tariff should always be based on design and BOQ. And expert consultants should be involved in preparing estimates against which competition is to be held. We have serious reservations on this as the following data and discussion would demonstrate including the facile way in which NEPRA has dealt with this issue and estimated the reference tariff in the first place which has encouraged the parties to quote exaggerated costs.

NEPRA based its reference tariff of 9.86 USD per tonne by extrapolating the AES data of 2009.It was a totally different type of offshore jetty in which 5-7 kms of Trestle had to be made under water foundations which is always very expensive. Also AES Jetty was designed for very large ships (CAPSIZE 150-200,000 DWT) carrying three times the coal carried by PANAMAX vessels. AES feasibility study is available in the offices of NEPRA and thus there is no excuse of not knowing the AES jetty details. No one could imagine NEPRA would take a highly ill-advised (to put it mildly) step of allocating the same costs as these would occur in AES and Gadani. The coal jetty being built as a part of the power plant is a much simpler and cheaper version.

Nearby, PIBT is in process of commissioning a common purpose Bulk Terminal which will provide coal handling services to other power plants. For strange reasons, NEPRA ignored data from this actual and comparable plant and used the inappropriate and irrelevant AES project data, which was also out of date. PIBT terminal, which is 4 times the size of Sino-Hydro Coal Jetty (16 MTPA vs 4.5 MTPA) is being built with a cost of 285 Million USD vs. 240 Million USD for Sino- Hydro. Reportedly, PIBT tariff is 4 USD per tonne (unless they have jacked it up seeing this example) as opposed to 9.86 USD per tonne for Sino Hydro coal project.

In India, coal terminals of twice the size of Sino Coal jetty have been made in 30-60 million USD range. In the kind of budgets approved by NEPRA, very large Coal Terminals in the range of 30-40 MTPA, 6 to 10 times larger in output and capacity, can be built. In India, typically coal terminal is USD 4.00 USD per ton (total including fixed and variable) as terminal tariff, out of which 40% (1.6 USD per tonne) is the capital charge rate and 2.5 USD per ton is the O&M cost. International rates of coal terminal are around 2-4 USD per tonne. Railway coal terminals tariff in India is 1 USD per tonne each on dispatch and receipt. These are all comparable numbers. However, NEPRA is awarding or has awarded 9.86 USD per ton as Capital Charge (240 Million USD CAPEX) of Jetty and 9.34 USD per ton as O&M costs, totaling a hefty 19.20 USD per tonne. The difference is too much to be left unexamined thoroughly, once again.

I am not sure, if dredging costs have been included in Jetty CAPEX. PIBT terminal and others internationally, let dredging included in Port charges to be directly collected by the port authority such as PQA directly from the Ship operator. Coal terminal charges are to be paid by the importer. Even on dredging, the quoted numbers are very high i.e. of 5.4 million tonnes. Dredging is one area, which is universally seen with a lot of skepticism and very rigorous mechanism should be in place through third-party involvement. Other similar projects at Port Qasim (DATANG EIA report) have estimated dredging requirements of only 1.5 million M^3. If the number of 5.4 M^3 is correct, then it is a problem of wrong siting. The record and proceedings that NEPRA has publically shared does not show any serious consideration or investigation by NEPRA with respect to either dredging or otherwise.

At first, a reference tariff was announced by NEPRA using data of an incomparable project and accurate determination was left for later. When the time came for detailed scrutiny, no serious investigation was made to improve upon it. A 240 Million USD investment being dealt with such flippancy is not expected of NEPRA. Clearly, transparency requirements have not been adequately met in the process of determination of cost-plus jetty CAPEX. Neither NEPRA and nor the company have published requisite information. The evaluation report circulated does not provide even salient data on cost elements and quantities, nor did NEPRA reveal the requisite data in this respect. NEPRA would be well advised to re-examine the case by involving independent experts and help remove confusion and controversy in this respect.

About Manjung Coal Power Plant Malaysia

Manjung 4 incorporates a variety of advanced technologies to operate at a main steam temperature of 600C and pressures above 4,000 psi. The boiler is a once-through, sliding-pressure, vertical-tube furnace wall, two-fireball, two-pass design equipped with Alstom's low-NOx tangential firing system. The vertical tube design incorporates rifled tubing and orifices to distribute fluid flow to the furnace wall tubes in proportion to tube heat absorption. Rifled tubing spins the water/steam mixture traveling within the boiler tubes, throwing the water onto the tube surface to aid cooling. Tubes in the center of the wall receive more heat and require more cooling, so proper fluid distribution reduces temperature differentials and stress within the furnace wall.

Sliding-pressure operation allows the plant to operate more efficiently at part load—the plant can ramp down to as low as 300 MW. This mode reduces the boiler pressure as load decreases, minimizing throttle valve energy losses and helping maintain high steam temperature to the turbine. This also reduces thermal stress during cycling.

The firing system uses two adjacent firing circles, creating two fireballs. Fuel and air nozzles can be tilted to control the reheat temperature without the need for spray desuperheating. A separated over-fire air (SOFA) wind box is located above the main wind box to optimize air staging and minimize NOx emissions. Each SOFA wind box contains six separate air nozzles, all of which are both vertically and horizontally adjustable. This highly adjustable combustion system allows for control of fuel and air conditions to suit many different types of coal, thus maximizing combustion efficiency while maintaining low emissions, particularly NOx.

The steam turbine is an Alstom STF100 unit equipped with one high-pressure turbine, one intermediate-pressure turbine, and two double-flow, low-pressure turbines. The turbine is paired with an Alstom GIGATOP two-pole generator. Though the design features necessary to handle such a wide variety of fuels result in lower overall efficiency compared to other ultra-supercritical plants, Manjung 4 still achieves nearly 40% efficiency, making it the most efficient coal-fired plant in Southeast Asia. Compared to Units 1–3, it generates 14% more power per metric ton of coal burned. Manjung 4's other operating parameters are shown in Table 1.

Like Units 1–3, Manjung 4 uses seawater flue gas desulfurization (FGD). About 20% of the seawater drawn for plant cooling is used in the FGD system, and the system can absorb more than 90% of SO_2 emissions. The FGD effluent is mixed with cooling water exiting the plant, treated to increase the dissolved oxygen, and returned to the ocean. Though this process results in a slight increase in sulfate content of the seawater, as well as lowering the pH, the effluent is well within environmental standards and creates no harmful by-products. SO_2 emissions are below 200 mg/Nm3 and NOx emissions are below 500 mg/Nm3.

Unlike the other units, however, which rely on electrostatic precipitators (ESPs) for particulate control, Manjung 4 uses an Alstom Optipulse pulse-jet fabric filter system that keeps particulate emissions below 50 mg/Nm3. Again, this was dictated by the wide variety of coal types the plant may be called upon to burn, which was too wide for optimal performance of an ESP.

Since coming online, the plant has exceeded expectations, achieving 2% higher output and heat rate than specified. The plant has given TNB a tool to maintain reliability despite continual fluctuations in the regional coal market. Further expansions of the Manjung site are under way. A fifth unit, the ultra-supercritical 1,000-MW Manjung Unit 5, is under construction by a consortium composed of Sumitomo and Daelim Industrial Co. and is expected to begin operations in late 2017.

26. New Thar Coal Tariff

Many people believe that Coal Power tariff has not been determined appropriately and mishandled by the stakeholders on government side. Even if competitive bidding option is not available under G-to-G bilateral projects, it is still possible to reach agreement on reasonable terms and prices. It should be made sure that vendor does not influence policy.

Thar coal tariff is due for renewal of tariff for which a public hearing was organized on 28th March, 2017. Ministry of Water and Power has untypically asked for downwards revision of the tariff by 20%. They had issued a similar concern regarding high renewable energy tariffs awarded by NEPRA. Earlier they, more often than not, used to ask for high tariffs. I have been raising the issues of balancing the incentives vs loss of market due to high prices, while in government and outside government. My book, Issues in Energy Policy, is replete with the undesirability of high tariffs. In this chapter, we will take a stock of the situation with respect to Thar coal Tariff in particular and coal tariff in general.

The major issue in case of Thar coal Tariff is the Rate of Return allowed at an IRR of 20% on Equity which is translated to 35.4 % RoE in operational years. Perhaps in the history of utility sector, nowhere in the world, such a high return has ever been allowed. The normal NEPRA rates for imported coal power plant tariff are 15% IRR on Equity translated to 24 % ROE. A 15% IRR on Equity compares well with others, say India, which offers the same return. However, that is in local currency and includes tax. Our currency on the average depreciates at 5% per year. In the local currency terms, NEPRA ROE rates translates to 20% IRR and 31% RoE. For Thar it would be as high as 25% IRR and 41% RoE in local currency terms, which actually matters. Nowhere in the world, have such high returns been offered. An incentive of 1-2 % would have been more than enough, if at all.

Engro manages to get better terms as a starting party. They got a very high tariff of 66 cents per   MMBtu in their RLNG terminal, while later parties, two years hence, and are getting 41 cents for the same. The same is happening in Thar coal. Good communication skills and stakeholder management is all that is possibly required to get a good deal.

I am not sure why NEPRA helped start a controversy on coal tariff by approving RoE instead of IRR on Equity. I assume they are equivalent. In ROE, returns are provided only in operational years, while IRR includes payments during construction. I would suggest return to IRR which is more versatile, understandable and comparable and perhaps even more transparent. Secondly, tariff determinations are a bit cryptic i.e. calculations are not easily verifiable. Presentation should improve providing major calculation steps. One has to develop his own model in order to fully comprehend the determination issued by NEPRA.

People used to yearn for local coal utilization expecting that it would be cheaper to do that and would save foreign exchange but this hasn't proven true for Pakistan. Thanks to the add-on costs awarded by NEPRA to imported coal power tariff such as about 20 USD per ton for port handling, there is some comparability to Thar coal prices vs imported one.

Thar coal price as proposed by Engro and approved by TCEB (Thar Coal Energy Board) in the initial 10 years is 50.69 USD per ton, which boils down to 3.84 USD per MMBtu. Imported coal presently is low priced due to oil price decrease partly at 69 USD per tonne as opposed to 2014 when reference tariff was awarded and international prices were 89-90 USD per tonne FOB. For those who may not be knowing, fuel costs are treated as pass-through. It does not matter, what were the prices when tariff was awarded, power producer is paid at current prices, which may be high or low, a fair arrangement taking out any gambling element that would have been there otherwise.

For Punjab based Coal Power Plants such as in Sahiwal, there would be another add-on in the form of rail transportation cost from Port to Site. Railways maintained that it is the GoP/ECC which determines railways tariff and have questioned NEPRA's jurisdiction in this respect, the latter was inclined to issue or approve transportation component as well. I am not sure if the issue has been settled and at what tariff. Similarly, Thar coal (mineral) part is not under purview of NEPRA.TCEB solely determines the coal price. These are tricky issues on which some deliberation is required. TCEB is a producer-linked agency and not a regulator to be independent.

Can there be reverse auction or competitive bidding? I am not sure if reverse auction can be there in coal power plants in Thar or otherwise. Competitive bidding can always take place, if one power plant is brought for tendering at a time allowing sufficient competition. However, in CPEC context, wherein bidding option is not available, what can be done? I have made some submissions in this respect in an earlier chapter in this book.

Is the CAPEX right is always a question. In the case of Solar and Wind power tariffs, which are simpler and have a lot of recent projects to draw upon, CAPEX issue is much simpler. In large plants such as coal and other thermal, the issue is much more complicated. If EPC costs are comparable, Non-EPC cost is an irritant and varies substantially from site to site and offers ample opportunities for those whose business model thrive on CAPEX padding and no equity. Even in EPC, there are many variants: thermal efficiency and coal quality affects CAPEX; environmental specifications are a major factor; SOx and Particulate matter controls are normally there in almost any power plant being installed these days but there are also the issues of their performance characteristics and stringency. Whether NOx controls are there or not can be a major variant affecting CAPEX.

Thailand has signed an agreement recently with ALSTOM/Marubeni for installation of a high efficiency (44-45%) coal power plant running on lignite coal. A 600 MW plant would have an EPC cost of 950 Euros which translates to 1.742 Million USD per MW. The plant would have very high environmental performance including SOx and Particulate control. I am not sure if NOx is included. In Malaysia, the CAPEX rate is 1.2 million USD per MW and in Indonesia 2.0 million USD per MW. In India, Alstom, Doosan, Siemens and Toshiba are supplying High Efficiency Super-critical and Ultra Super-critical coal power plants at rates of 0.9 to 1.1 million USD per MW. For Thar Coal Type Lignite based power plants, the rate is around 0.9 million USD per MW.

One can either be sure under bidding or a feasibility estimate done by experts. Both are unavailable in Pakistan. NEPRA thinks that collecting a motely crowd of semi-experts is enough for correct determination. Bureaucracy cannot be cured of the know-all syndrome. Or it is penny-wise pound-foolish approach? A small amount of money paid to experts (international) may save hundreds of millions of USD. Overall, it appears that CAPEX of imported coal power is about right. One is not very clear about the specs, especially, on environmental equipment.

In India, on pithead coal power plants, NTPC India has been able to maintain coal power tariff at 4.5 cents on the average, although non-pithead plants produce at about 6 cents.50% of coal tariff is fixed cost and about 50% is fuel and variable cost. Lignite based electricity is more expensive, which is about 6.6 cents, as revealed by CERC-NLC determination.

NEPRA took international prices of coal at an average of around 89 USD per ton FOB which adding transportation and handling of another 40 USD per ton becomes 129 USD per ton or 5.763 USD per MMBtu. International coal prices have been coming down and were almost halved in 2015 to 40 USD per ton. Prices have now recovered to 69 USD per tonne. At current prices, if reference tariff is computed now, it would be at 109 USD per Ton or 4.8323 USD per MMBtu.

The major advantage for Thar coal appears to be that there is no transportation costs which would otherwise add about 40 USD per tonne extra costs in transportation and handling. Thar Coal is of low CV of about 6000 Btu per lb as against normal sub-bituminous coal of 10,253 Btu per lb. Thar Coal price compares favorably; 3.8402 USD per MMBtu with international/Indonesian reference prices of 5.763 USD per MMBtu. In that case why Thar Electricity tariff is almost equal to imported coal tariff? Perhaps, high RoE is responsible. Thar coal's Electricity may become cheaper, if ROE is corrected and thermal efficiency is enhanced.

NEPRA tariff is silent about power plant specs. It only specifies on performance measure such as Thermal efficiency. It is silent on environmental performance, relying on EIAs. One is not sure what is being actually supplied and installed in ongoing coal power plant projects. At COD, they only measure Heat rate and electrical supply variables. I hope, either NEPRA or EIA procedures provide for the measurement of environmental performance at COD as normally environmental issues receive least attention. But is this case, consumers are actually paying for environmental equipment in the form of tariff and any leniency or deviation should not be acceptable.

One is not sure as to what is the logic of issuing tariff for Thar coal transported to Port Qasim. Has feasibility been done for proving the transportability, which cannot be done until some Thar coal is actually available, which would take another six months or one year. There have been counteracting news and discussions regarding transportability of Thar coal which has to be tested against actual field-testing. Some time back, an agreement was reached with the Sindh government that a percentage of Thar coal would be used in imported coal power plants. That has perhaps been forgotten, for, neither the NEPRA tariff mentions about it, nor any effort has been done by the IPPs to design their boiler furnaces accordingly. Thar coal may perhaps be used only when the foreign exchange issue makes it necessary. To be fair, this could not have been done due to the short time that was available to handle the unknown Thar coal.

Way Forward

I have made the following recommendations to NEPRA; 1) Thar Coal IRR on Equity be brought in line with other coal power plants i.e. 15%; 2) KIBOR margin should be reduced to 1 %; extra margin to be awarded on proof; 3) An international study should be commissioned on LIBOR margin to bring down LIBOR margin if possible;4) Coal Handling Terminal charges and its CAPEX be reconsidered ,although it is related with

Table 36 Comparative Tariff: Thar vs. Imported Coal

	Units	Imported Coal	Thar Coal	India-lignite	India Sub-Bitumen
Capacity	MW	1099	330	500	660
CAPEX	Million	1483.7	497.7		
CAPEX per MW	Million USD/MW	1.3500	1.5082	0.9	0.9
Thermal Efficiency	%	40	37	35	35.5
Tariff	Usc/kWh	8.0139	8.5015	**6.6**	4.5
Energy Cost	Usc/kWh	4.9161	3.5411		2.5252
CPP	Usc/kWh	3.0978	4.9604		1.9748
ROE	%	24.5	30.65		15.5
Coal price	USD/ton	129	50.69	34.68	30
Coal price	USD per MMBtu	5.7633	3.84	2.735	2.6273
CV	Btu/lb	10253	6000	5780	6800
Capacity factor	%	82	85	85	85
Exchange Rate		97	97	66.6	66.66

Source: Compiled by the author from various sources such as NEPRA-Pakistan, CERC and NLC India

imported coal;5) USC and SC technology on Lignite is now competitively available from many sources and thermal efficiency figures of (44-45%) may be accommodated within the available CAPEX in the existing determination if possible. High efficiency comes at a cost. Chinese should be consulted, if those kind of thermal efficiency figures are available with them on Lignite of 600 MW and over.

Concluding, proposals for reducing Thar coal tariff should not be understood as being against Thar. On the contrary, a balanced approach is being argued. High prices increase producer's incentive. Low prices means buyer's incentive. Solar and Wind promoters in this country kept lobbying for higher prices and got them, but resultantly, very few projects could be installed. History of other countries has indicated that market and investment expand at low prices. Thar promoters have to adopt a market sense. There is no monopoly of any energy source .If there was any, it has been broken by the imported coal which one would like to discourage, which compares quite unfavorably with Thar Coal power tariff of Rs.8.5 per kWh. Imported coal tariff should have come down automatically with reduction in international coal prices by 25-50%. Thar coal should be kept competitive with imported coal. And solar prices have come down to 6 cents as against a coal tariff of 4.5 cents in India. Secondly, circular debt can only be eliminated, if the gap between cost of production and consumer tariff is reduced along with eliminating Theft. All will have to participate proportionally in this effort.

Post-script

New Thar coal Tariff has been announced by NEPRA on July 27 2017(These lines are being added when the manuscript has been completed; some repetition may be there, due to this factor). It would be valid for the next 2 years or 5000 MW whichever occurs earlier (presumably). The earlier Tariff remained applicable for 2640 MW comprising of EngroPower (2x330=660 MW), Thal Nova (330 MW), Thar Energy (330 MW), Thar Coal Block I (1320 MW).One keeps hearing of the first two projects and not much about the latter two. We will apprise the readers here the main features of this tariff.

The new thing about the tariff is that it has introduced two Tariff systems one based on wet (water) cooling project and the other on Dry (Air cooling).Normally wet cooling is used and all of the coal power plants are based on water cooling. Most of us are skeptic of dry technology on account of its lower efficiency, higher CAPEX and higher OPEX.I must confess that I was among one of the sceptics. However the data provided by the Chinese (Shanghai Electric) who have the right experience on this had a different opinion. They have argued that, in fact, there was not such a formidable difference in case of dry cooling so as to be totally excluded from consideration. This in fact was proved as we see in the results of the tariff determination released by NEPRA and which is the subject of discussion here. There is a 2% difference in efficiency (lower), and a difference of 0.11 cents in cost of generation. Following are the levellised Tariff figures:

USc7.1318/kWh for two units Air cooling foreign currency financing (FE)
USC 7.2228 /kWh for Single Units Air cooling FE
Usc 7.2275/kWh for double unit FE WET cooling
Usc 7.3356/kWh for Single Unit FE WET cooling

This is a good news for Thar coal, as it was being thought that there would be an upper limit on as to how much electricity can be produced from Thar coal. The resource is large but there are water availability issues being the main constraints. The upper limit was conceived and calculated by some experts to be 10,000 MW. Now with the advent of Air cooling technology and not being uneconomic as well, one could look forward to much higher production levels. It is a separate matter that due to worldwide opposition to coal and some of the local policies,(the latter we will discuss later herein), Thar coal production and financing may be discouraged.

A major issue in case of Thar coal tariff was the allowed IRR of 20% on equity which translated to 35.4% ROE in operational years. NEPRA has reduced IRR to 18% and has dropped the term of ROE which had created a lot of confusion. GoS kept insisting on an IRR of 20% thinking that it would be good for Thar coal. But many, including this scribe, thought that the opposite was true. The argument of the latter group is that Thar coal has to compete with other sources of energy, even within Sindh, of Wind and Solar energy. There has been a tremendous drop in the international and regional prices of Wind and Solar power, reaching a level of 4-5 cents. NEPRA has rightly decided on competitive bidding which will help resolve the local controversy spread by the vested interest in favour of higher and higher tariff. An IRR of 20% was really high; it was justified for the risk of first few projects and there were times when local interest rates were as high as 14% and a margin of 6% was defended. Local interest rates today are of the order of 6% and added another 6% margin, IRR comes out to be 12%.Infact, and Zorlu has submitted a tariff application for a solar power project that is based on 12% IRR. Investments are driven by political considerations. Some of the western countries investors would not be attracted by even 20% as has been proved already in most of the power projects including Thar. And Chinese would have considered 12-14% to be good enough and would have been attracted. There isn't much market to sell coal power plants any way. It is a mistake to award an IRR of 18% under pressure from GoS who haven't given much thought to the issues as have been explained here. The real beneficiaries and vested interested that appears to be behind higher IRR are the local parties for whom nothing would be enough. But as they will see, it would hurt them and hurt us all.

The idea of differential IRRs (returns) is not tenable anymore. GoS nationalist demand higher returns on Thar coal, and KPK nationalist also demand matching returns for Hydro Power. In fact one of the NEPRA members from KPK wrote a note of dissent on a hydro tariff case, using just this logic. Let every source of energy compete on its own and let the buyer decide, how much to buy and when, based on (Least Cost Generation Planning) and not through arbitrary and random political decisions coming from Lahore.

Another good thing introduced by NEPRA in the new Thar Power tariff is reduction of interest rate margin from 4.5% in case of Sino-sire fee applications to 4.00 % in case where Sino-sure insurance rates are not applicable. This was long overdue; der ayad durust ayad. When hard times come to pay in future, one may make a case for downwards reduction in other cases retrospectively. It was in fact unreasonable to charge a heavy Sino-Sure insurance of 7% on debt and charge normal commercial rates as well. In fact under the non-competitive bidding cases as most of the Chinese CPEC projects are, there is a case of negotiated lower rates of financing under G-to-G arrangements. The logic being that there is higher CAPEX in such situations. Let us be honest, the upfront tariffs are not that independent. In India, Lignite coal tariff is Rs 6.6 current, which in levellised terms may still be lower.

Higher Tariff of Thar Energy Tariff is also due to apparently higher coal prices dictated by Thar Coal Energy Board (TCEB).I suppose they are still using IRR of 20%. There hasn't been any public hearing about it. TCEB does not believe in public consultations and prefers working out such important rates within themselves, although I note quite some respectable names on the committee that has been entrusted by TCEB this important task. Is provincial autonomy such a bad thing so as to preclude transparency and public consultation and oversight? Some reform is required in this respect? If unilateralism prevails, then tomorrow, KPK may ask exclusive role in pricing its electricity.

In the first year, the Thar coal cost/price as expressed in NEPRA determination is 14.75 USD per ton Variable, and an unduly high fixed components of USD 56.43 to give a total of 71.18 USD per ton. In later years, variable component goes down to USD 10.64 per tonne and fixed cost of 19.41 USD per ton to give a total of 30.05 USD per ton of total cost. Accordingly levellised cost is 46.50 USD per ton. Engro keeps saying that once, an optimum production level of its mine is reached, the production cost would come down. On the other hand, one keeps hearing of new coal mines investments, all of them submitting similar cost schedules as indicated here. If all of such new proposals keep coming and approved, how will that purported optimum level would reached. Under similar stripping ratio conditions in Central Europe, Lignite is being sold at a price of 20 USD per tonne or even lower. In the U.S., it is as low as 10 USD per tonne. In India, it is IRs. 1500(24 USD) per ton. Reportedly, international consultants had been hired by TCEB based on which, such pricing policy has been prepared.one wouldn't mind ,some good royalty payments going to GoS which it would hopefully spend on social sector, especially, in Thar. One would be advised to look into the possibility of some kind of competition among coal mining companies in order to bring down the cost.

Why are Thar Coal prices so high as compared to elsewhere, as noted above? Firstly, regulated tariffs are almost always high. If competitive bidding is adopted, it is hoped that the prices would come down. We have provided a comparative table of the three coal tariffs that have been issued by TCEB from which, it can be easily seen the rising cost trend, while the opposite ought to have been the case. As mentioned elsewhere, what is the logic of entertaining other coal tariff applications, when optimum levels of existing tariff commitments have not been achieved yet, as indicated by EngroGen figures? Tying up, coal mining projects with power appears to be the reason, which should have been delinked by now. Old technology, such as Shovel and Trucks use might also be the reasons for higher costs. One cannot move millions of tons of overburden with shovels and trucks. It takes more time and energy. In India and Europe, where most of the Lignite mining is being done, Bucket Wheel Excavators have been used. India introduced BWEs long time back in 1960s on Nvyeli Lignite Mines. Two more such machines have been acquired by NLC India. Admittedly, these are expensive and heavy on upfront cash, the unit product cost is lower than Shovel and Truck as are being used currently. In Engro tariff determination, 161 million USD of Diesel consumption has been shown as CAPEX item. TCEB should seriously consider the submissions that have been made here to bring down Thar coal costs.

The cash flow based tariff calculation model as applicable to Electricity tariff has been applied to coal/Lignite mining as well. It is unprecedented and not found almost anywhere in the world. Not only that, very high cost of production of coal has resulted due to this and even more importantly, lignite cost as high as 71.18 USD/ton in the first year results jacking the first year tariff very high and almost unaffordable. Let us take, Oil and Gas model, as applied in this country and as where. Does any, oil or gas producer require the buyer to pay for his debt repayment and RoE. There is an oil or gas pricing formula according to which buyer is supposed to pay. Same applies to coal. It depends on the oil/gas/coal producer to finance his CAPEX and OPEX, as he deems fit. The integrated model of Engro being the first project is being pushed too far. This is causing fragmentation preventing scale economies. Power generation has to be separated from Coal production. Let there be coal mining companies producing and selling coal to the IPPs. There would be operational and transparency issues when the integrated companies, e.g., Engro, would be selling coal to other IPPs. One of the solutions could be to establish Sindh Coal ala Coal India or Gujarat Mineral Development Corporation .Sindh coal may invite/tender coal mining companies to install coal mines, independent of IPPs, buy coal from coal mining companies and sell coal at a composite price or may designate an IPP to buy coal, as is being done in case of gas.

Table 37: Comparative Tariff Determinations of Thar Coal Mining Projects

		Engro	SinoSSRL	Sindh Carbon
Capacity	MTPA	6.5	7	4
Capacity Power	MW	660,1320 MW	2x660=1320	2x300=600 MW
CAPEX	MnUSD	839.67	1049	672.56
Unit CAPEX	USD/ton	129.18	149.86	168.14
Stripping Ratio			7.86	8.21
Water Withdrawal	Cusecs	30		81
Production Tariff	USD/ton	13.64	26.94	29.01
Capacity Charge	USD/ton	32.49	17.45	31.11
Total Tariff(levellised)	USD/ton	46.133	44.36	60.23
Equity Returns	USD/ton	7.8387	8.81	18.96
incremental CAPEX	USD/ton	40.278		
Royalty	%	7.5	7.5	7.5
ROE	%	20	20	20

Source: TCEB

In a typical mining contract, following is usually done; a minimum off-take is guaranteed, some advance payment is made, there is some pricing formula or a price agreed to with an escalation clause providing for inflation. This avoids, unduly high payment/cash liabilities as are being incurred in the current system adopted by TCEB/NEPRA. A competitive bidding can be easily organized along these lines. Chinese government may also be requested to cooperate and support such bidding among its own companies, after all its image is also at stake during all this. Daily, some critical articles appear in the press narrating stories of Sri Lanka and other African countries who are having debt servicing and cash-flow issues. We can avoid the pitfalls which have been suffered by others. All stakeholders would benefit, if these submissions are considered.

27. Environment and Coal Power

Coal power has acquired a bad image, perhaps less due to its pollution dimension and more due to the CO_2 emissions causing climate change, of which Pakistan has been projected to be one of the ten worst victims. Despite, criticism of coal, hundreds of coal power plant keep operating in the world with more than 100,000 MW installed in India alone. At this moment, we are planning to produce about 5,000 MW in coastal areas, 5,000 MW in Thar and may be 1,000 to 2,000 MW in Punjab. In the medium to long term, this number might be doubled. Even with this much addition, our contribution to GHG and climate change will be insignificant both in absolute or per capita terms. However, we have to be careful about the pollution issue, which would damage no one but ourselves more directly and immediately. Abundant technologies are available, which if applied, can produce clean electricity even on coal as many nations have started doing it, including China.

Health and environmental consequences of large-scale coal power production should not be ignored. Four main emissions, namely particulate matters, SO_x, NOx and mercury compounds, are released into the air by burning coal. Bronchitis, eye diseases, lung problems, heart attacks, mental disorders and stunted growth of children are some of the health problems that have been linked to these emissions. Moreover, NOx and SOx are washed down due to rain causing acid rains, which affect agricultural productivity. Mercury pollutes water bodies and finds its way into human beings through food chain. Pakistan has signed Minimata Convention in 2013, which aims at controlling or even almost phasing out mercury, as mercury pollution spreads far beyond national boundaries. However, responsible action can minimize these adverse health and environmental impacts. Pollutants can be arrested at source by installing appropriate equipment. High efficiency can reduce CO_2 production, which causes greenhouse effect.

Pollutants generated by Coal Combustion

Particulate matter can be controlled by passing the exhaust gases through an ESP (Electrostatic Precipitator), which catches carbon and dust particles by static electricity phenomenon. ESP equipment has to be highly efficient with an efficiency rating of 95-99

%. Moreover, the devices have to be maintained rigorously, otherwise these lose efficiency resulting in dust emissions.

Sulphur and its compounds are absorbed by passing flu (chimney) gases through calcium and useful gypsum is produced as a result of the chemical reactions. Gypsum has uses in building and agriculture sector. In technical terms this is called FGD (Flue Gas Desulfurization). In alternative processes, Sea water, being alkaline, is used as a scrubbing fluid producing sulfate salts (normally benign) that are drained into the sea along with the used sea water. No reagents or chemicals are required and a very small OPEX is required in the form of energy consumption in pumping water. It is the cheapest and more affordable FGD process appropriate for coastal regions where seawater is available. One wonders why such an inhibition is there in the coal IPPs of Karachi.

NOx is controlled either by installing special burners that produce less NOx. This is a cheaper and less effective method. More effective but expensive methods are using ammonia absorbing flue gases in NH_3 and breaking down NOx into nitrogen and oxygen or through catalytic converters, as are installed in cars these days to absorb NOx. The process is known as Selective Catalytic Reduction (SCR) whereby NOx is converted, with the aid of a catalyst into diatomic nitrogen N_2 and water (H_2O).

Mercury is often absorbed in Flue Gas Desulphurization (FGD), although there are specialized processes for it. Particulate controls are being installed in all parts of the world, rich and poor. However, FGDs are less common, NOx treatment even lesser, and mercury control, the least practiced. Even in the U.S., as recent as 2000, almost 50% of coal power plants did not have SOx control. However, these days, no coal power plant can be installed without FGDs in any developed country. Existing plants either have to be closed down or retrofitted with these devices. Mercury control does not generally require specialized equipment, and pollution control equipment employed for particulate matter (ESP), SOx (FGD) and NOx (SCR).

Enforcement of Pollutant Controls in Pakistan

There appears to be confusion over installation of pollution control equipment in the coal power plants that are being installed in Pakistan. In generation licenses applications, all the technologies (ESP, FGD, and SCR) are required to be installed. However, in the Environmental Impact Assessments (EIAs) the pollution equipment appears to have been watered down. EIAs of Karachi and Sahiwal projects are not publicly available contrary to the general practice. It appears that only ESPs would be installed for particulate control. NOCs issued are also equivocal and general. For example, Sindh Environmental Protection Agency (SEPA) mentions the requirements in general terms as required under SEQS rules. PEPA (Punjab Environmental Agency) has issued No Objection Certificates (NOCs) mentioning ESP and Mercury control (even mentioning Minimata convention membership of Pakistan) conspicuously avoiding SOx (FGD) and NOx (SCR) control equipment specifically. However, one wonders, how mercury would be controlled without these devices. NEPRA accepts NOCs of EPAs without giving much thought about it, and issues tariff without discussing or providing for environmental costs. NEPRA does not even examine the contents of generation license while issuing approvals, as it appears. Some interaction ought to be there between the two sets of agencies on such issues, as we will discuss later.

In Punjab, cities like Sahiwal where a coal power plant has been set up, and in congested and polluted cities like Karachi, SOx and NOx controls are a must. How can one install large coal power plants (1,300 MW) in these locations without such controls? It is quite amazing that our EIA community (consultants and EPAs) manages to provide recommendations and conclusions convenient to the client. If the client does not like to install environmental control equipment like FGD and NOx controls, they prove with their complicated models that it is not required. There can be other ways of avoiding compliance by assuming/promising the usage of low sulfur coal irrespective of whether it `is ever used actually and that cheaper high Sulfur coal would not be used, as has happened in case of High sulfur fuel Oil (HSFO) that has been used in this country against all the environmental rules rather unabashedly. Avoidance of environmental equipment is being done under these provisions. It is surprising, how, EIAs can prove that no FGD would be required and low Sulfur can meet the environmental standards. No two EIAs match. For example, EIA of DATANG (700 MW proposed to be installed at PQA) concludes that Sea Water FGD would have to be installed even with the questionable assumption of using low sulfur coal (0.5%).The larger Coal power plants of 1,100 MW are equivocal about it, to say the least.

On the other hand, JICA and ADB have got EIAs done for their coal power plants in Jamshoro area, which require FGD and accordingly FGDs are being installed there. Responsible institutions like World Bank, Asian Development Bank, IFC etc. therefore mandate installation of environmental control equipment and do not approve of hiding behind loopholes.

Environmental controls, however, cost money both in terms of higher Capex and Opex. But such cost is much less than the cost of bad health and adverse environment. Besides, there are treaty obligations. FGD used to be quite expensive, when market was smaller but its prices seem to have come down. ALSTOM is installing FGD in India in 25 Million Euros for 500-600 MW coal power plants. If we add 50-100 Million USD in our 1300 MW coal power plant, it may not be a bad deal. After all, NEPRA has approved a jetty capex of 300 Million USD recently. A separate add-on component may have to be considered for SOx and NOx removal equipment, if and where it is installed. These may not be required for projects installed in Thar area, due to its remoteness, low population and no agriculture. In addition, CFB boilers employed in Thar power plants absorb sulfur during the burning process. This would add to the tariff and can be controversial, as many people think that already coal power tariff is high compared to other countries.

We can learn from bad experience of India. Environmental performance of the power plants, especially, of the coal power plants is very pitiable in India. Reportedly, more than 100,000 people die in India due to particulate matter ensuing from the coal power plants. Until, recently, there were no environmental standards for coal power plants. Only, a limited number of coal power plants installed have only particulate matter controls. SOx and NOx controls are not there at all. We do not have to repeat their example but should rather learn from it. However, India has introduced corrective environmental legislation on coal power plants, which has initiated a round of FGD installations and other controls through retrofitting projects. New coal plants will have to install such equipment from the beginning.

Chinese, on the other hand, have the wherewithal to provide add-on environmental equipment. The traditional image of China insensitive to pollution should go away. Today in China, more stringent standards are being adopted than there are in the U.S. and many deviant coal plants are being closed. The new plants that are being installed are highly environmentally compliant and the cost is reasonable as well. If seriousness is shown from our side, they would be able to implement those schemes here as well. However, in this short time that was available, not much could have been done. Let us look forward to have some movement in this respect, once basic equipment has been installed.

Environmental measures cost money but the alternatives of not taking these measures cost much more. International bodies have judged Karachi and Lahore to be among top ten most polluted cities of the world. Smog in winters in Punjab blocking road and air transportation for weeks are easy reminder of the realities.

Clean electricity can be produced within our existing tariff, if tariff system is reformed reducing excessive payments such as 20 % IRR on Thar Coal projects and 17% on imported coal, and doing other refinements, as I have suggested elsewhere. These excessive payments can be directed towards installing much needed environmental equipment. As for reducing carbon footprint, high efficiency (45%) power plants enable one to do that and such enhancement pays for itself in the form of more energy production or less fuel consumption. Fortunately, Ministry of Water and Power has asked NEPRA to reduce high returns of developers, which may create space for adding environmental cost in the tariff. A stitch in time saves nine. Installing pollution control equipment may cost much less in the beginning than retrofitting later. Concluding, the relevant agencies should get together under the leadership of the Ministry of Water and Power to remove the policy vacuum without waiting for public protestation that may ensue due to environmental deterioration in the relevant areas even though it would be affecting future governments.

Concentrations of Air Pollutants in Cities of Pakistan

An analysis of the available data from 2007 to 2010 (measured and compiled by Pak EPA and the allied international agencies) shows very high concentrations of fine particulate matter (PM2.5)—measured in micrograms per cubic meter ($\mu g/m^3$)—in Lahore (143 µg/m3), Karachi (88 µg/m3), Peshawar (71 µg/m3), Islamabad (61 µg/m3), and Quetta (49 µg/m3). Most likely, the high value concentrations reported in this analysis would have been even higher if the monitoring instruments had been working all the time. Particulate matter of less than 1 micron (PM1) and PM10 measurements were not available. Low data coverage (average of 17% for the five cities) partially affected the PM2.5 measurements.

Lahore had highest sulfur dioxide concentrations. The analysis of the 2007–10 time series on SO2 confirmed that Lahore was the city with the highest concentrations (74 ± 48 µg/m3), with maximum daily values of 309 µg/m3. Other cities presented very high values of SO2: Quetta (54 ± 26 µg/m3), Peshawar (39 ± 34 µg/m3), and Karachi (34 ± 34 µg/m3). Overall, SO2 values were found to be increasing over the course of the study period (2007–10).

Table 38: WHO Guidelines for Ambient Air Quality (mg/M^3)

Mean Values	SOx	NOx	PM2.5	PM10
Annual		40	10	20
24-hrs	20		25	50
1-hr		200		
10 minutes	500			500

Source: IEA

Pakistan's nitrogen dioxide concentrations are slightly above WHO guidelines. The annual nitrogen dioxide (NO2) concentrations derived from the 48-hour data revealed that the current levels in the country are slightly higher than the WHO air quality guideline value of 40 µg/m3, with the highest concentrations in Peshawar (52 ± 21 µg/m3), Islamabad (49 ± 28 µg/m3), Lahore (49 ± 25 µg/m3), and Karachi (46 ± 15 µg/m3). Concentrations were somewhat lower in Quetta (37 ± 15 µg/m3). Results from an analysis of data from 2007 to 2011 show that concentrations of ozone (O3) and CO were well within the WHO guidelines as given in Table (Source: Clean Air Pakistan by Asif Shujaa, ex-DG Pak EPA).

One could readily see from the Table 30, Emission Standards of Coal Power Plants, that most countries have toughened these standards. Most interesting is the fact that the new Chinese standards are more stringent than those of Germany. And new Indian standards are also either more stringent or equal to the standards of the advanced countries. Old India, South African, Thai and Indonesian standards were comparable, however, all of them have improved upon their standards and more is expected to follow. One is, however, surprised to look at a very lax standard on NOx of Australia: maybe there is an error.

Table 39 Emission Standards of Coal Power Plants: New vs. Old and Comparative Environmental standards of various power plants (mg/M3)

Selected countries' emission standards for NOx, SOx and PM from coal-fired power plants							
Country	Time period	NOx, µg/m^3		SOx, µg/m^3		PM, µg/m^3	
		Existing	New	Existing	New	Existing	New
Australia			800		200		80
China	hourly	100	50	200/50	35	30/20	10
Germany	daily	200	150	200	150	20	10
India		600/300	100	600/200	100	100/50	30
Indonesia		850	750	750	750	150	100
Japan		410	200		200	100	50
South Africa	continuously	1100	750	3500	500	100	50
Thailand		820	410	2002	515	180	80
USA	Daily	135	95.3	185	136	18.5	12.3
EU IED	Continuously	200	150	200	150	20	10

Source: IEA

	Parameter	Port Qasim CPP	Sahiwal CPP	Sahiwal CPP Actual		HUBCO CPP (	Engro CPP (units)
				Unit 1	Unit 2		
1	PM	32.17	31.69	4.95	14.22	< 40	11.6
2	NO x	450	300	176.97	119.92	< 398	-
3	SO x	17.5	18.54	106.29	97.44	-	33.8

Source: presentation Director Coal, PPIB

Post Script: At the time of going to the press, it was revealed by PPIB that Environmental controls have in fact been installed at the two coal power plants at Sahiwal and Port Qasim. Data for Hubco and Engro which are under implementation has been taken from their EIA as indicated by PPIB. It appears that there is problem with NOx .It is not clear whether SCR have been installed, otherwise, NOx ratings would not have been that high.

28. Underground Coal Gasification

There is an R&D project under which Thar Coal is being gasified underground (hence the name Underground Coal Gasification-UCG) for electricity production and possibly for producing other products such as fertilizers and chemicals. The project has managed to attract quite some attention as well as controversy. There are proponents and opponents having strong and weak arguments defending their positions. We will take stock of the project and try to guide the readers as to the merits and demerits of the arguments and the project itself. We will try to explain as to what the project is; what has been achieved and what has not been achieved and cannot possibly achieved; what is the international status of the project and what are the prospects of commercial interest of IPPs and developers to adopt UCG as a process and where does our R^D project stands in this respect; what are the options to benefit from the R&D efforts that have been done up-to-now. We will try to explore answers to some of the questions and issues in this short space.

UCG is an old process of gasifying coal mostly developed in Russia and Uzbekistan. Coal Gas was used in homes and elsewhere before World War II. There are three processes through which coal can be used productively including UCG. One is of digging the earth and getting the coal out and burn it in boilers of the power plants and produce electricity. Currently, in Pakistan, there is a major move to utilize both imported and Thar coal in the CPEC programme and outside of it. The other is of gasifying coal, once coal is brought onto the surface trough underground mining or open-pit mining. Coal is half-burnt(under sub-stoichiometric conditions, plainly speaking, with less air or oxygen) in closed pressure vessels and a mixture of Carbon-Mono-Oxide,CO_2, and Hydrogen is produced which has a calorific value .The process of Coal Gasification in vessels over-ground is well established for many years. Fertilizers and chemicals are being produced, especially, in China and South Africa, using this process. Big and credible Companies like GE, Texaco, Conco-Phillips, Lurgi and Shell have developed and perfected their processes. Thus per-se, there is no doubt as to the gasification of coal and even of its economics.

However, UCG is not commercially established and could not proceed beyond R&D projects in several countries. There was a surge of interest since the last decade, which unfortunately could not be sustained for a variety of reasons. And under the same surge, Pakistan also launched this programme under the courageous and enthusiastic leadership of the famous nuclear and missile scientist Dr. Mubarakmand. What I can testify is that he has been able to successfully gasify gas and producing electricity out of it and is doing that currently and in that the stated R&D objectives have been achieved. However, for such projects, third-party validation is usually required for approving more resources or closing down the project or continuing with it in some form or the other. According to the Minister of Planning Prof Ahsan Iqbal, the same has not been provided. I can perhaps also testify that this is true as well. Third Party validation is a must and should be financed by the Planning Commission, as resources are not left with the UCG project and also because the independence requirements dictate so. Bulldozing the next stage of the project through press campaigns is not a very good idea and it should have been avoided.

Many people including myself appreciate the efforts of Dr. Samar Mubarakmand and his able team. Sincere and persistent R&D efforts are in short supply in Pakistan and those who do it must be appreciated and encouraged. However, there are issues and questions on the road ahead. Dr. Samar insists that further resources be committed to commercialize his project to a capacity of 50-100 MW as against a few MW that he claims he is producing and consuming itself. What skeptics argue is that if some net electricity is being produced, why it is not being supplied to nearby villages or road lighting or to the nearby projects. Is it not the case, God-forbid, that no net energy is being produced or the efficiency is below 10%, not worthwhile to be pursued? Data and answers should have come out of the project in public domain in writing. Skeptics also argue that nobody in the world is pursuing UCG and that if UCG and our R&D is worthwhile, why don't IPPs and foreign developers take interest in it. Unfortunately, Coal is going down and companies are withdrawing from coal. In older times, companies could invest for the hack of it. Scientists like ours cannot go possibly beyond R&D. There are unresolved issues such as environmental impact, sustaining an interrupted and sustained combustion, coal wastage and efficiency (it is apprehended that only 10% of coal may be utilized in the process and the rest being rendered unutilized for good resulting in simple waste and thus poor economics). There are issues like this that have to be answered by the Third Party Validation. Mere enthusiasm is not enough of an input. Besides, commercializing requires a variety of resources and technologies which may not be possibly mobilized or available within our national system. Financing of the R&D itself, perhaps was a major enterprise and all decision makers and stakeholders have to be appreciated in this respect.

On the other hand abandoning the project to be depreciated and stolen away in the Thar Desert may not be a good idea as well. There should be new technologies and skills that should be absorbed, retained and documented. It should serve as a focal point and a nucleus for allied research. The project should be fenced and given under may be Geological Survey of Pakistan or PCSIR and some commercial laboratories be built within it. PCSIR itself may be the right institution to conduct third party validation and to suggest as to how to go ahead in consultation with the team of Dr. Samar. The latter's chances of getting resources for the above-mentioned appear to be more than a capital intensive appropriation for a questionable commercialization project.

29. Industrial and Agricultural uses of Thar Coal

In this Chapter, we will discuss uses of Lignite other than large scale electricity generation which we have focused upon mostly throughout this book. There are many other uses of Lignite such as industrial and space heating, production of humic acid, soil conditioners and in making additives used in Oil and Gas drilling.

Lignite has been used in industries for space heating and industrial boilers and in rural areas in Germany and is still being used. Briquette production used to be 49.39 million tons in 1989 which however, has come down to only 1.54 million tonnes in the year 2016. Dry pulverized coal, however, has maintained its market share and presence at around 4.71 million tonnes per year. These are significant quantities. By comparison, Pakistan's coal imports mostly being used in cement plants, are to the tune of more than 5 million USD per year.

Both Briquettes and Dry Pulverization improve calorific value, reduce volatile and waste material, reduce pollutant gases and improve transportability. Lignite is highly volatile and can be instantly ignited under certain weather conditions. Lignite, in its raw form cannot be transported beyond about 100 kms, as many studies indicate. Hence, Lignite, as-mined, has not entered into international trade via sea or even land routes.

The processes are fairly simple; drying and grinding (pulverization) in one case and in the other making Briquettes after drying by applying mechanical pressure as is used in Bricks making. The difference is that Briquettes are smaller.

Despite significant success in adding electricity generation capacity, there are issues in primary energy supplies. Petroleum is largely imported. Gas resources are being depleted and expensive LNG (despite halving of the international prices) had to be imported. Local export industry complains of falling competitiveness due to lower gas prices in Bangladesh and lower electricity prices in India. Although, this is a temporary situation. Bangladesh is fast moving towards a high energy cost and price regime. Its gas resources are also being depleted and has started launching RLNG import projects which will make its electricity and gas expensive. In Pakistan, cost of electricity generation is expected to come down when expensive furnace oil is replaced by cheaper coal and more efficient RLNG based NGCC plants.

Thus, availability of cheaper primary fuel to industry and rural areas would give fillip to local industry and improve living conditions in rural areas. Processed (Briquettes or Dried Pulverized) lignite can replace imported coal for cement plants, used in Brick Kilns, and can be burnt in industrial boilers of textile and other industries. Only minor conversion is required to convert from gas and oil. Rice mill boilers and other rural industries can also be converted to this fuel. It should save more than 50% of fuel cost of industries. In cold rural areas like FATA, AJK and Gilgit and Baltistan, Chitral and elsewhere, institutional buildings can utilize boiler based space heating using this fuel.

Even in the power sector, there may be a requirement of processed Lignite fuel. Imported coal power plants that are being installed are utilizing PC boiler technology which permit a narrow range of coal specifications. These plants would not be able to use Thar's lignite coal except in a very small percentage of mixing with imported coal. It is quite a debatable question that alternative CFBC technology should have been used in these imported coal projects, as in that kind of boiler, the possibility of using Thar lignite would have been there. The Mine-mouth power plants that are being installed by Engro under CPEC are, however, based on CFBC technology.

Processed Lignite can, however, be used in PC Boilers of imported coal projects such as Sahiwal and Port Qasim. Fortunately, technology is now available to process lignite to produce dry pulverized coal that can be fired in PC Boilers. USDOE has spent quite some resources on it. It should be possible to have transfer of technology through USAID. After all, pollution characteristics are improved under this kind of processing. Drying Lignite, although, has become a well-known technique and is being widely used practice and there is no patent or special rights on it. It is fairly simple. Waste heat of Power plants can be used to dry the Lignite which can be pulverized using conventional means. This also means that Thar Coal Power Plants installed in Thar can also generate additional revenue producing this Dry lignite which, as we have mentioned elsewhere, has a large market in areas outside Thar.

In Pakistan, PCSIR has conducted research on producing a Furnace-Oil like product by grinding coal and mixing water in it along with some additive. Similar research has been successfully done elsewhere. PCSIR should repeat this R&D on Lignite, one Thar Lignite is available. All such efforts have been done looking for a cheaper alternative to expensive Furnace oil.

It is a misnomer, I should add, that CFBC Boilers as are being installed in Thar do not require Sox and NOx controls. While it is true that significant amount of Sulphur is removed as a solid waste during Lignite combustion with calcium, the Sulphur removal is not enough to meet the stringent environmental standards of today. It is a separate question that in the remote areas of Thar, is such an environmental stringency required? Environmentalists may have a different point of view but they may have hard time convincing people and the industry.

Some people may raise eyebrows on a suggestion to advance proposals on utilization of Coal/lignite in the manner that has been done here, arguing that why should it be done when everybody is leaving it. My argument is that we are a poor and underdeveloped country facing many challenges. We should adopt a mid-of-the road position on the issue of Climate change. Our exports are either falling or not increasing at a desirable rate and imports increasing. At the stage of development that we are in we should try to use all the available energy resources. Solar and Wind are electricity generation resources. We need primary thermal energy as well for industrial uses. Gas resources are depleting and imports of LNG may not be sustainable.

Agricultural and other uses of Lignite

Lignite can be used to make Humic Acid, which is a fertilizer and can be used to make a compound called Leonardite which has many uses. Wikipedia has following to say on it:

> "**Leonardite** is a soft waxy, black or brown, shiny, vitreous **mineraloid** that is easily soluble in alkaline solutions. It is an **oxidation** product of **lignite**, associated with near-surface **mining**. It is a rich source of **humic acid** (up to 90%) and is used as a **soil conditioner**, as a stabilizer for ion-exchange resins in water treatment, in the remediation of **polluted** environments and as a **drilling** additive. It was named after **A. G. Leonard**, first director of the North Dakota Geological Survey, in recognition of his work on these deposits.[It is used to condition soils either by applying it directly to the land, or by providing a source of humic acid or **potassium humate** for application.
>
> Leonardite can be added directly to soils to reduce the take-up of metals by plants in contaminated ground, particularly when combined with **compost**. **Leonardite** is used to stabilize and thin fluids used in the drilling of wells for hydrocarbon exploration/extraction and **geothermal** drilling. It was first employed extensively during **World War II** when **quebracho** became difficult to obtain. It has very good temperature stability and prevents solidification of lime muds near 150 °C."

It is pleasing to note that our universities and researchers are doing research on the subject. GoP may consider launching and promoting projects to expand the uses of Lignite as has been mentioned.

Figure 14 Processes of Pulverized Lignite Burner

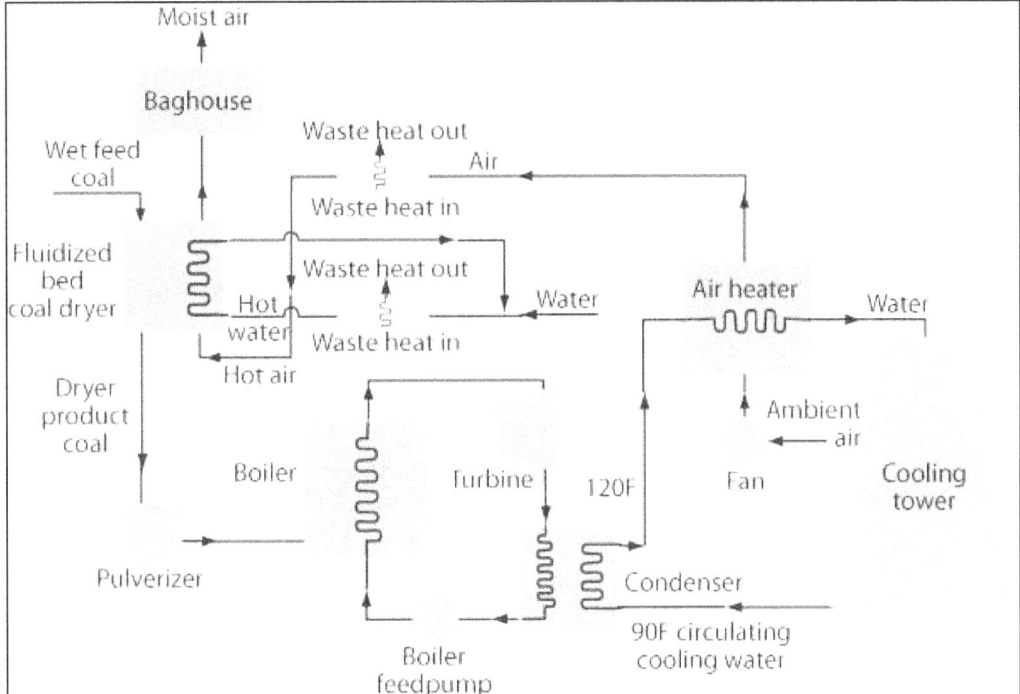

Source: USDOE, Powermag.

PART IV: ELECTRICITY

30. How much electricity do we need?

Energy demand and Supply issue has become a very important yet complicated question in the wake of heavy investment in the energy sector that is being done. There are sceptics and pessimists who are warning that the debt crisis would emerge due to heavy CPEC investments and that economy would ultimately suffer under balance-of-payment issues. Within the government departments, there is now a question circulating around as to how much of energy capacity is enough?

Conceptual Issues

Energy has many elements or components; electricity, Gas, Petrol, Diesel, Kerosene etc. Petroleum products deserve lesser attention because these can be easily imported as and when required and long-term capacity development issues such as storage are not that critical. Gas and Electricity, generally, cannot be imported that easily. Transport infrastructure such as pipelines, specialised terminals and transmission lines are required and take a long time to install. In this section, we will focus on these two only, namely: electricity and gas demand estimates and projections of the future, say 2025 and 2035.

It has been established through many empirical studies, that there is a two way causality between Energy consumption and growth of economy; economic growth spurs demand of energy and energy supplies and consumption drive economic growth. This is also understood as a common sense observation but has been verified through empirical evidence and studies.

To cut the story short, in order to be able to project the energy-electricity demand/consumption, you should know what the growth rate of economy was in the past and that of energy and that what the quantitative link of the two growth rates was; economists call it Energy-GDP elasticity. Secondly, one should know what would be the rate of growth of economy over the long run(2025 and 2035).There are disagreements whether the current year forecast rate of GDP would be achievable; then what to talk of these years. Thirdly, you should be able to know or intelligently guess, what quantitative changes would occur in future relationship of GDP and Energy. It used to take 131 kg of oil consumption in the past to produce a GDP of 1000 USD (PPP-2011) and in 2015, it required only 105 kg to produce that much GDP i.e. with advancement and ever increasing energy conservation and efficiency possibilities, lesser and lesser energy is required. Thus Energy-GDP elasticity would be less that the past and then the guess-work that what would be the rate of such adjustment, year-on-year or as a long term average.

To make matter worse, there are people, who would legitimately argue that price has an intuitional and empirical causality with energy consumption. It is even more difficult to forecast energy prices inflation in real terms, even if determination of past linkages is accomplished with difficulty. In order not to make the job of demand projection to be impossible, let us restrict ourselves to Energy-GDP relationship. In nutshell, we will explore as to what was the energy-GDP elasticity in the past and what would it be in future (guess-work) and what would be the growth rate of the economy over an extended period of time (ambition vs. realities and accidents?).If in the past, a GDP growth rate of 4% required an increase of 5% in energy consumption, then at what rate of growth in GDP (say of 7%-ambitious or 5% pessimist or realist), what rate of growth in energy consumption can be expected. The reader would understand, why our bureaucracy did not give much attention to the question, although studies have been done in the past on this question. When, it is all guesswork then why bother? The issue is common for all economic forecasting. Economics is a dismal science; economists first develop a forecast, which comes out to be wrong and then they spend rest of the time why they went wrong.

We have mentioned earlier that there is a difference of opinion as to a GDP growth rate of 5.2% would be achieved, then what to talk of future. But one should have ambition, if not a target, in order to be able to manage his affairs and set directions. CPEC institute has reportedly made a statement that due to CPEC, Pakistan's growth rate would become 7.5%. Perhaps the basis is that energy shortages had an impact of 2% reduction in economies' growth. CPEC would eliminate the energy shortage and thus add 2% to growth of 5-5.5% in the economy. Vision-2025 predicts or ambitions a GDP growth rate of 7-8% by the year 2025, as I understand. It would be a good achievement, if we are able to achieve that in the context of difficulties that we face. India has managed to achieve a growth rate of 8-10 % in the last decade and continues to do so. In earlier years, India used to have growth rate of 3-4% so persistently, that a low growth rate was termed to be Hindu growth rate. And now, India is having consistent high growth. Will such a change come in Pakistan is a million dollar question? And thus what growth rate of energy would be more appropriate for the future is then a 2 million dollar question instead of one.

There is no consensus in Pakistan yet, as to what has been the energy-GDP elasticity: JICA has estimated an elasticity of 1.1, while some researchers have estimated a number of 1.5.There is a need of evaluating all the divergent studies and arrive at a consensus number and its forecast for the future. In India, a GDP elasticity of 0.8 is being assumed for future planning purposes. In Europe, this number varies from 0.5-0.6.I would have a judgement around 1.0 for Pakistan for planning purposes, meaning there by that if economy grows at a rate of 5%, energy would also grow at a rate of 5% and if the economy grows at a rate of 7% p.a., energy demand / consumption would grow at 7% p.a.

Ironically, there was never such a question or situation. Investment was always less than the demand. But now doubts are being expressed whether the investment would be more than the demand required. If CPEC is not there, the same situation of under-investment would occur. IMF and other IFIs claim that Pakistan's economy cannot afford a FDI more than 2 billion USD per year. Then a searching question is, can Pakistan's economy in the view of IMF declared limitation be ever able to invest the required amount in energy sector. Or late Dr. Mehboobul Haq was correct when he said that some load shedding may be optimal, as peak supply is very expensive.

In this hour of need and load-shedding, the answer to the question posed in the title of this chapter would be as much as one can supply. But the future needs are uncertain. After an investment in 10,000 MW, how much further capacity do we need, so that this time we are able to plan and implement accordingly and avoid a crisis of the nature and scope that we are facing today? Hence, the pressing issue is how much electricity do we need and how much fuel or primary energy will be required to generate this quantum of electricity and to meet the other thermal energy needs of homes, industry and transportation. Fortunately, petrol, diesel and furnace oil are importable and now even gas is being imported in the form of LNG. Electricity is not imported generally, and requires as much long cycle in importing as local generation infrastructure does.

Pakistan's electricity consumption, compared to other countries, as given in the table below, is very low indicating lagging development resulting in unemployment, poverty and conflict. Pakistan has a per capita consumption of only 472 kWh per year, as opposed to 805 of India and 1,699 of Egypt. I had thought that the three countries are almost equally poor. It may be the case, but certainly Pakistan has lagged in development as is indicated by almost the lowest possible electricity consumption per capita in the region. Perhaps, we are only better than Afghanistan, Bangladesh, Mali and Chad etc. I am scared to look for the data of these countries, lest I get disappointed even further. Ignorance is a bliss. But let this not lead into an unwise stampede under the lure of CPEC. We should make realistic assumptions, projections and plans, otherwise, we would further slide into the kind of darkness we find ourselves in.

Table 40 Salient Data on Economies and Electricity consumption of Selected Countries- 2014

Country	Population	GDP (Nominal)	GDP (PPP)	Electricity Consumption	Electricity Consumption per Capita	GDP/Capita (PPP)
	Million	Billion USD	Billion USD	TWh	kWh/capita	USD
Pakistan	185	206	832	87	472	4,498
Egypt	90	238	882	152	1,699	9,855
Turkey	77	871	1,392	220	2,870	18,172
Thailand	68	383	996	174	2,566	14,705
Indonesia	254	942	2,501	207	814	9,846
Malaysia	30	314	717	139	4,646	23,980
Philippines	99	251	646	70	706	6,516
India	1,295	2,196	6,902	1,012	805	5,330
South Africa	54	411	659	229	4,240	12,204
China	1,364	8,230	16,841	5,357	3,927	12,347
U.S	319	16,157	16,157	4,137	12,962	50,649
Russia	144	1,676	3,220	950	6,603	22,361
France	66	2,729	2,407	460	6,955	36,376
Germany	81	3,624	3,438	570	7,035	18,584

Source: IEA-World Energy Statistics, 2014

Many attempts have been made to forecast the electricity demand based on which investment recommendations were provided. Most of these went astray, but the reason for crisis is not the inaccuracy of forecasts but the implementation. Hopes were tied to attract IPP investments, which never came in the volume that was required and local resources were either not there or were not applied. IFIs were more focused on Smart Meters, CASA, renewables and price and subsidy reforms. Domestic political issues prevented Thar Coal and KalaBagh dam to be implemented, while external issues like India's opposition to Bhasha prevented IFIs to finance Bhasha. It is only CPEC's grape-wine, which has energized the energy sector and that we are talking of how much would be enough lest we invest in overcapacity and incur debt that we cannot service. Thus the issue of demand forecast has become extremely important.

Two major studies have been done on the subject of electricity demand and supply: study done by SNC-Lavalin in 2008 and JICA's load forecast study in 2016. SNC-Lavalin study is based essentially on regression modeling that attempts to correlate and develop a model interrelating electricity demand with GDP, energy price and number of customers etc. JICA, on the other hand, utilizes Least-Cost Generation Planning software. It deals with electricity and as well as primary energy demand issues. Both the studies disaggregate demand into its sectoral components such as domestic, commercial, industry and agriculture.

It should be noted that both the studies have been done by outsiders with NTDC only providing logistics and may be data support for the external consultants. The Planning Commission made a local effort in 2006 through mobilization of local resources and another one by the same institution in 2010 through ADB's technical assistance and by acquiring external consultants. The study did not yield results and was terminated due to lack of funds. Planning Commission is now reviving the same initiative with the assistance of the USAID. The difference between Planning Commission's approach and MoWP/NTDC approach has been that NTDC restricted itself to electricity, while Planning Commission attempted to develop integrated energy model and thereby the forecast of all energy elements including electricity. PC-ADB project used a specialized modeling language called MARKAL-TIMES developed in the IEA system for policy analysis purposes.

What is to be noted is that not only that our own institutions did not or could not undertake these studies, they did not or could not even update those studies. At least, there is no record that such an effort has been done and shared government-wide. Academia did attempt something but the output was less than desired. ADB consultants in the Planning Commission study had involved academia and expensive software licenses were provided to a number of leading universities. Except for UET Taxila, no university showed sustained interest. It is important to do this kind of self-criticism so that such inadequacies and practices are not repeated again.

All models require updating. Its coefficients change with new data and structural changes and it is only through such updating that local and organizational capacity develops. NTDC (SNC Lavalin) study contained all the wherewithal for updating. Nothing was withheld unlike JICA study wherein consultants did not reveal their mathematics in the report but discussed it orally in presentations.

Most models involve GDP growth and energy price as exogenous (independent) variables. Some make assumptions about changes in the structure of the economy. One can idealize about having as many variables included as possible. But the problem is that reliable data is not generally available. And still more difficult is the forecast of the chosen independent variables. What use is the model containing many variables, when reliable data is not available? Energy and GDP relationships can yield sufficient reliability for data reasons. Even, in the case of GDP one would not know what would be the growth rate. In the kind of political and social circumstances, it is hard to predict, yet some planning and programming is required. It is often said that economists develop forecasts which often go wrong and they spend rest of the time as to why did their forecast went wrong, so as to be able to do a better job next time. It is due to this reasons that possibly much attention has not been given to this side. I have earlier explained that the problems are not due to lack of accurate or rigorous forecasts, it is due to implementation. Everybody knows and knew that electricity and energy demand increases every day, especially, in the high population growth and low level of electrification that we are in.

Many imponderables were there and more have joined recently. The issue of Energy conservation and Efficiency and the emergence of Renewable energy such as Solar are affecting demand and other associated variables like T&D losses, in innumerable ways. Look at what has been done by CFL lamps vis-à-vis incandescent bulbs and now inverter ACs are doing.

It would be safe to assume the Energy-GDP elasticity to be unity i.e.in plain language, energy/electricity demand would grow at the same rate as that of GDP growth. The forecast for the new financial year of GDP growth, as recently announced by the Planning Commission, would be 5%, which many skeptics are doubting. In any case, it may be reasonable that mid-term growth rate may remain to be 5%, given the political instability and the approach of elections. In this circumstance, a growth rate of 4-5% energy/electricity consumption demand may be taken as reasonable.

Demand Side Estimates

In the early stage of development, energy demand grows faster than that the growth of the economy. As the economy treads on development trajectory, energy consumption slows down. For most of the low performing economies of Africa, an Energy-GDP elasticity of 1.5 may be normal. For India, it has been projected at 0.8 and even lower after being at unity 1.0 for many years (Planning Commission of India, erstwhile). For most European countries, it is less than or around 0.5(EON study). All due to conservation and energy efficiency expectations. For Pakistan, the number may lie in between. There are methodology issues. I have examine many academic papers in this respect with almost every paper giving a different number. However, if there is any convergence, it is about unity.

Table 41 Comparison of JICA and SNC-LAVALIN Studies: Peak Demand Forecasts (MW)

Scenario	2025		2035	
	JICA	LAVLIN	JICA	LAVLIN
Base case/Normal	48,399	44,199	79,958	99,012
Low	NA	36,954	NA	67,285
High Case	55,382	51,870	102,338	144,190

Source: JICA and SNC-Lavalin Reports

1. JICA 's High case of 102,338 MW peak demand in 2035 compares very well with the Normal/Base Case forecast of SNC-Lavalin of 99,012 MW.

2. There is close correlation in the peak demand forecasts for the year 2025. JICA estimate for 2025 base case is 48,399 MW, while SNC Lavalin estimate for the same year is 44,199 MW. High case estimate of JICA for 2025 is 55,382 MW, while for SNC-Lavalin it is 51,870 MW. There isn't a huge difference between Base case and the High case estimates for the year 2025. However, for the year 2035, there is a vast difference in the estimates of the two studies, in both cases: 79,958 MW JICA estimate vs 99,012 MW of SNC-Lavalin and 102,338 MW of JICA estimate vs 144,190 MW of SNC-Lavalin for the High case estimates of 2035.

3. Following net conclusion can be made;

- Peak Demand for the Year 2025 can be taken as 45,000-50000 MW.
- Peak Demand for the Year 2035 can be taken as 90,000-100,000 MW.

This conclusion correlates with demand estimates based on electricity demand growth rate of 7% per annum or that the demand doubles every 10 years. Assuming a unitary value of elasticity of Electricity demand vs GDP, this would imply an economy growth rate of 7% p.a. on the average for the whole forecast period. This may be taken as an upbeat and optimistic projection. The lowest growth rate scenario of electricity sector would yield a value of 4% p.a. and roughly the same level of growth rate in the economy

(GDP). Over the last 20 years (1995-2015), the growth has been at an average compound rate of 4.13 %. If the circumstances and policies repeat, the same can happen again. Supply momentum due to CPEC, however, has reached a level that the same cannot happen again in near to midterm future, except the consequences of a debt-laden economy and its bad management conjures such a situation. This kind of forecast (of around 4% growth in demand) gives us an electricity demand of 35,582 MW in 2025 and 52,663 MW in 2035, which are rather low figures. Having said that, it appears that most likely values would be in between the low and high growth rates of 4% and 7% in electricity demand that may be consistent with a growth rate of 5- 7% in the economy, given the continued role of conservation and energy efficiency.

Under this demand scenario, there does not appear to be any justification of new project sanction beyond those that are already in pipeline. No more imported coal power plants, whether in Punjab or in Sindh, are required. There is sufficient number of Thar Coal projects in the pipeline, which should be enough. Similarly, there is no scope for any further RLNG plant for the time being. Bhasha, which is already approved and even partly implemented in the sense that Rs 100 Billion have already been spent on it, must be brought to construction stage and it should be completed by 2025. Bhasha is a life and death issue as discussed in another chapter of this book. We require it more for water purposes than for electricity as for electricity there are many other viable options.

Some renewable energy projects 1,000-2,000 MW may be inducted, provided reasonable prices result as a consequence of the proposed auction. Government should be careful in entertaining the hype of high growth lobby, which wants to get its projects approved through CPEC. They will not suffer under take or pay contracts. Only consumers, people and government would suffer. If they are so sure of high demand, let them come under take and pay contracts. Government appears to be in the same frame of mind as a result of JCC deliberation that has been reported in the press. I am not privy to any special information.

Over and above the proposed projects, in the mid-term, GoP should pay attention to the T&D sector. A lot of skepticism is being expressed on the ability of this sector to deliver. Already, power cannot be evacuated from a number of wind power projects. Additional investment would be required in this sector requiring financial allocation and foreign investments. No to further generation projects should also be understood in this perspective.

Table 42 SNC-Lavalin Study: Electrical Peak Demand Forecast 2035

	2014-15	2024-25	2034-35
NORMAL FORECAST			
PEPCO	21,171	38,781	85,600
KESC	3,007	5,824	14,322
Pakistan	23,958	44,199	99,012
LOW FORECAST			
PEPCO	20,863	32,473	58,366
KESC	2,960	5,824	14,322
Pakistan	23,606	36,954	67,285
HIGH FORECAST			
PEPCO	21,265	45,450	124,273
KESC	3,021	6,895	21,241
Pakistan	24,066	51,870	144,190

Source: NTDC

Table 43 JICA Study: Electricity Peak Demand Forecast up to 2035

	2015-16	2025	2035
BASE CASE			
North	16,728	29,546	49,071
South	8,649	14,801	24,794
KHI	3,222	5,459	8,423
Pakistan	26,499	**48,399**	**79,958**
HIGH CASE			
North	16,775	33,809	62,806
South	8,673	16,936	31,734
KHI	3,222	6,247	10,781
Pakistan	26,499	**55,382**	**102,338**

Source: JICA Study LCGP NTDC 2016

Table 44 Demand and Supply Projections up to 2035

	Growth Rate	Existing Demand 2016	Demand 2025	Demand 2035	Comments
	%	MW	MW	MW	
Low Growth-I	4	23,000	32,736	48,458	Low
Medium Growth-II	5	23,000	35,681	58,120	Expected
Normal Growth III	6	23,000	38,858	69,589	High
High Growth-IV	7	23,000	42,285	83,180	Very High
Very High growth-V	10	23,000	54,233	140,666	Impossible
Depreciation			10,000	15,000	
Conservation ^Efficiency			10,000	15,000	
Net Depreciation + Efficiency			-	-	
Existing Dependable Capacity		18,000			
Gap/New Capacity Required		5,000			
Gap/New Capacity Required I	4		14,736	15,721	Affordable
Gap/New capacity Required II	5		17,681	22,439	Reasonable
Gap/New Capacity Required III	6		20,858	30,731	High
Gap/New Capacity Required IV	7		24,285	40,896	Very High
Gap/New Capacity Required-V	10		36,233	86,433	Impossible

Source: Compiled by the Author

The Supply Side: do we need more?

It may be noted in the adjoining tables that projects of a total capacity of 34,489 MW (CAPEX 80,986 Million USD) are already approved and in the pipeline. If one adds to it, the existing Capacity of 25,000 MW and excludes depreciation of 5,000 MW of capacity till 2025, it adds up to a total capacity by 2025 of 54,489 MW. It is quite evident that it more than matches the expected demand. There may be some uncertainties, say, about nuclear K2 and K3. Some under estimation may be there, e.g. ORACLE coal plant and other mining projects. One or two projects may be dropped but would be substituted by others, e.g. Rahimyar Khan Coal power plants of 1300 MW. Bhasha is already under implementation. Advanced negotiations are already underway on CPEC financing for this important project. This is almost certainly the most important project in this schedule. Some more solar and wind power projects of 2,000 MW capacity may be added either as substitution and may come from CPEC and non-CPEC sources.

10,000 MW of this capacity is already at an advanced stage of completion, to be commissioned in 2017-18. Contrary to the popular perception, 7000 MW of this 10,000 MW is non-CPEC investment of RLNG plants, Tarbela IV and V, Neelum Jehlum, PAEC nuclear C3 and C4 etc. Dasu is an approved World Bank financed project. Most NTDC and DISCO investments are financed by ADB. The question that arises, is, do we really need more supply other than some substitutions and modifications? It is already 80 billion USD plus. Fortunately, most of it is arranged financing of one sort or the other. Should we worry or celebrate? It depends which side of the political divide, you are. After a period of under-investment of the last two decades, this heavy influx was inevitable. Luckily, the money is available, whether one can mentally and physically afford it would continue to be debated for a long time.

In Table 35, we have provided demand projections under 5 different scenarios ranging from 4% Rate of Growth (RoG) thru 5%,6%,7% and 10 %. 5% has been the historical RoG of electricity generation. It is this low RoG that has caused the current electricity crisis. 5% can be an appropriate RoG for demand expansion, which may be compatible with a GDP growth rate of 5%, assuming a likely value of 1 for Electricity-GDP electricity. 5% has been the growth rate of the economy for most years over the last three to four decades. If JICA's estimate of electricity-GDP elasticity of 1.1 is accepted, electricity demand growth rate would be 5.5% at 5% growth rate of the economy. Only rarely has the economy grown at a rate of 7%, although successive governments have aimed at it. We can, therefore, take 7% RoG as an optimistic, yet feasible rate of growth of electricity demand. It however may be taken as an upper limit. This would give us a doubling period of 10 years.

We have assumed the demand level of 23,000 MW in 2016, with a load-shedding of 5,000 MW and a maximum peak generation of 18,000 MW. There might be some variation in this assumption. Further demand has been estimated based on this level. Existing installed capacity by the end 2016 was 25,838 MW (effective Capacity 18,000 MW). New capacity in the period 2013-2025 would be installed to the tune of 30,291 MW. By 2025, installed capacity with depreciation of 5000 MW would become 51129 MW .In the period 2025-35, it is projected that another 37200 MW will be installed.

After depreciation of all inefficient oil and gas based capacity of 14630 MW, the total NET installed capacity would reach a level of 73,699 MW. From the demand projections, we know that at 6% growth in demand, the Power demand in the year 2025 would be 39,000 MW and in the year 2035, it would be around 70,000 MW. And at 7% growth in demand, the demand in 2035 would be around 83,000 MW. Thus the supply estimates that we have prepared, match with a growth rate projections that lie between 6 to 7%.

Table 45 Summary of New Additions and Net Total Capacity by 2035

	2013-2025	2025-2035	Existing	Total
Coal	8,520	10,000	100	18,620
Hydro	10,489	10,000	6,902	27,391
Solar	1,000	5,000	400	6,400
Wind	1,000	5,000	306	6,306
LNG	5,600	3,600		9,200
GAS		-8,842	8,842	-
Furnace Oil		-5,788	5,788	-
Nuclear	1,100	1,100	1,000	3,200
Total	27,709	34,700	23,338	85,747
KESC	2,582	2,500	2,500	7,582
Total	30,291	37,200	25,838	93,329
Depreciation	-5,000	-14,630		-19,630
Net Total	25,291	22,570	25,838	73,699
Capacity/yr.	3,365.67	3,720		

Source: Planning Commission, MWP

Table 46 List of Electrical Power Projects

		CAPACITY (MW)	CAPEX (MnUSD)
A	CPEC-Approved		
1	2×660MW Coal-fired Power Plants at Port Qasim Karachi	1,320	1,980
2	Suki Kinari Hydropower Station, Naran ,Khyber Pakhtunkhwa	870	1,802

3	Sahiwal 2x660MW Coal-fired Power Plant, Punjab	1,320	1,600
4	Engro Thar Block II 2×330MW Coal fired Power Plant	1,320	2,000
	TEL 1×330MW ,ThalNova 1×330MW		
5	Surface mine in block II of Thar Coal field		1,470
6	Hydro China Dawood Wind Farm(Gharo, Thatta)	50	125
7	Hydro China Dawood Wind Farm(Gharo, Thatta)	50	125
8	Imported Coal Based Power Project at Gwadar,	300	600
9	Quaid-e-Azam 1000MW Solar Park (Bahawalpur)	1,000	1,215
10	UEP Wind Farm (Jhimpir, Thatta)	100	250
11	Sachal Wind Farm (Jhimpir, Thatta)	50	134
12	SSRL Thar Coal Block-I 7.8mtpa &SEC	1,320	2,000
13	Karot Hydropower Station	720	1,420
14	Three Gorges Second Wind Power Project	100	150
15	CPHGC Coal-fired Power Plant, Hub,Balochistan	1,320	1,940
16	Matiari to Lahore ±660kV HVDC Transmission Line Project		1,500
17	Matiari (Port Qasim) —Faisalabad Transmission Line Project,		1,500
18	Thar Mine Mouth Oracle Power Plant	1,320	2,000
B	**ACTIVELY PROMOTED PROJECTS**		
19	Kohala Hydel Project, AJK	1,100	2,397
20	Rahimyar khan imported fuel Power Plant	1,320	1,600
	TOTAL	13,580	25,808
C	**OTHER CHINESE INVESTMENT**		
21	PAEC-K2	1,100	5,000
22	PAEC-K3	1,100	5,000

23	BHASHA	4,500	11,250
	Sub-Total(21-23)	6,700	21,250
	Total Chinese Investment(CPEC plus others)	20,280	47,058
D	Non-Chinese Investment		
22	DASU	4,320	11,000
23	RLNG-I	1,200	1,200
24	RLNG-II	1,200	1,200
25	RLNG-III	1,200	1,200
26	NEELUM JEHLUM HYDRO	969	2,500
27	TARBELA-IV	1,410	914
28	TARBELA-V	1,410	914
29	ADB Jamshoro Coal fired Power Plant	600	900
29	SUB-TOTAL(22-28)	12,309	19,828
30	**GRAND TOTAL(A-D)**	**32,589**	**66,886**
E	**OTHER INVESTMENTS**		
	NTDC and DISCO(ongoing and Future projects)		10,000
	KESC	2,500	5,000
	Sub-Total(31-32)	2,500	15,000
	GRAND TOTAL(A-E)	35,089	81,886
F	EXISTING	25,000	
	Total		
	Depreciation(till 2025)	5,000	
	NET TOTAL CAPACITY-2025	55,089	

Source: Basic data from Planning Commission, MoWP

31. Load Shedding and Transparency

Intense heat and heavy load shedding has compelled me to write something about it as to why load shedding occurs and how it could have been reduced if not eliminated altogether. The issue is quite controversial and all kinds of advice and accusations are available of which we would like to take a stock of.

In all the advice and proposals that are normally given, Transparency of the type that is required is generally not included. We will posit some important proposals in this respect as well. We will take up the issue of load shedding first and discuss transparency later.

An obvious answer is that supply is less than the demand. But what is supply and what is demand? There is something called the Installed Capacity and there is another called de-Rated capacity because plants get old and lose their original capacity to produce for a variety of reasons. Then there is actually utilized capacity. Actual (de-rated capacity) may not always be utilized properly due to Poor maintenance and planning.

Surprisingly, there may be other reasons not to produce what one can produce. For example, it is alleged and perhaps correctly that K-Electric (and even the Ministry of Water and Power) does not run its furnace oil-based plants for financial reasons. NEPRA has fined them, but to no avail.

K-Electric has de-rated capacity of 2,093 megawatts and with purchase from local IPPs (350MW) and exports from Pepco system (650MW), the total supply capability adds up to 3,093MW. People say there should be no load shedding. K-Electric claims there is no load shedding in the paying regions where theft and receivables are under control. However, the fact is that there is load shedding.

K-Electric has a peculiar tariff system. It has been given a constant price tariff with adjustments for inflation. In this system, it is in its benefit not to run inefficient plants or the ones where fuel cost is more than what is provided in the constant price. Also furnace oil is purchased on cash and gas perhaps can be bought on credit or adjusted with receivables. However, one wonders that with oil prices 50% lower, is furnace oil issue still valid and instrumental in explaining the case. For a precise answer, one has to really audit the accounting data and do some scenario analysis. I am not sure if NEPRA has ever appointed some clever and professional auditors and consultants to find out the truth.

Even if K-Electric may be browbeaten into utilizing its furnace oil capacity, how about the Ministry of Water and Power? It is being accused of the same sin albeit perhaps for slightly different reason.

The government says the real dependable capacity is around 18,000MW, which it produces in summers. In winters, hydroelectric power is limited to 1,000 MW instead of 6,000MW that is generated in summers and thus 11,000MW is generated during winters.

Load shedding in summers is 8,000MW and in winters it is 5,000MW. Demand, however, has been increasing by about 1,000MW per year over the last few years, making the demand-supply balance or lack of it even worse.

Figure 15 Demand and Supply Situation January to November 2015

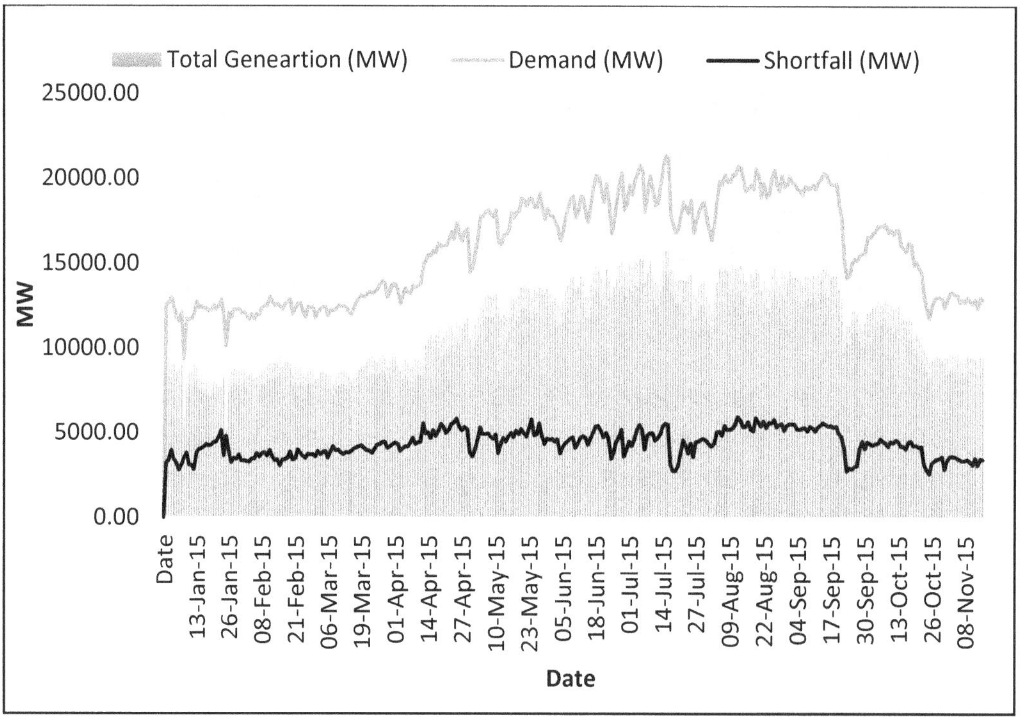

Source: JICA

Only 1,000MW has been added since 2013. Sometimes early heat makes the balance even worse as is the case these days in mid-April. The problem with electricity is that it comes or goes in lumps. It cannot be done yearly. It takes about three to five years to install a power plant of 1,000MW.

About 10 or more power plants with a total of 10,000MW capacity are at various but advanced stage of development and this capacity can only come online when plants are completed and commissioned in all respects.

One part less and the power plant would not be on. In some cases, part of the plants can be commissioned in advance as has been done in Bhikki where gas turbines have been installed and commissioned while steam turbines would be commissioned with a gap of some six months. Similarly, Neelum-Jhelum hydro may begin with one or two turbines.

Even if there are some problems in this package of 10,000MW whether in generation, transmission or distribution, it can be reasonably hoped that the electricity crisis would be largely over and may be difficulties would remain that are always there in life. Cheaper electricity from LNG and coal should also help reduce the cost of generation.

Under-Utilization of Capacity

In table 15, we have provided salient and condensed data on installed and de-rated capacity, annual generation and Capacity Utilization factor (CUF) of three categories namely hydro (which is almost all in public sector), Genco (which are in public sector and are thermal) and IPPs (which are mostly thermal and in private sector except for Chashma nuclear plants, which are government-run through PAEC). The table has been compiled out of the power system statistics (2015-16) released by the NTDC. Some power plants may have been left out for a variety of reasons.

Table 47 Aggregated Capacity Utilization Factors

	Installed Capacity (MW)	De-Rated Capacity (MW)	Generation (GWh)	Capacity Utilization factor (%)
Total	22,959	20,358	95,573	47.52
IPPs	10,061	8,960	47,667	54.08
GENCOs	5,852	4,352	14,755	38.00
Hydro	7,046	7,046	33,151	53.71

Source: Compiled by Author using data from Power System Statistics 2015-16

The most important column is of capacity utilization. Average total CUF is 47.52%, which is rather low, but at such an aggregate level may not serve the diagnostic purposes. Hydroelectric power sector has a CUF of 53.71%, which is normal for such plants. We can also understand the low CUF of 38% for Genco, which could have been at 50% under better and more efficient management. Gencos have also suffered from the lack of fuel availability. However, what is not fathomable is the low CUF of IPPs, which should have a high CUF of around 70-80%. Around 10,000MW of IPPs' installed capacity should give 7,000-8,000MW.

Adjoining Table 39 further confirms the claim that plants were under-utilized last year. These 8 power plants, given in the table, had an installed capacity of 8,260 MW and a de-rated capacity of 7,259 MW but only 2,917 MW was utilized. The shortfall was 4,737 MW (28,975 GWh). Thus, the capacity utilization factor of these plants was 42%. It has been alleged that these plants could have run at higher (80%) CF level, but for lack of supply of fuel, this could not have been done. Purportedly, fuel /furnace oil was not supplied because it was expensive and would have caused higher cost of generation, resulting in increased subsidy payments and generation of circular debt.

Thus, critics rightly argue that the ministry is not letting all the generating capacity being utilized and more costly oil-fired plants are not utilized as much as they should. At $50 per barrel, it is highly unlikely that this would help improve finances and reduce circular debt.

Table 48 Major Power Plants and their Utilization in 2015-16

Power Plant	Installed Capacity	De-rated capacity	Generation	Capacity Factor	Plant	Fuel
	MW	MW	GWh	%		
Jamshoro	850	700	3600	48.35	ST	FO
Guddu	640	425	149	2.66	ST/NGCC	GAS/FO
Guddu 5-13	1,762	1,762	3,409	22.09	NGCC	Gas
Muzaffargarh	1,350	1,130	5,148	43.53		GAS/FO
Kot Addu	1,639	1,342	6,583	45.85	GT/NGCC	GAS/FO/HSD
HUBCO	1,292	1,200	7,547	66.68	ST	FO

AES Lalpir	362	350	1,818	57.33	ST	FO
AES Pakgen	365	350	824	25.77	ST	FO
Total	8,260	7,259	29,078	40.19		
Actual	8,260	2,917	32,049	42.01		
Potential	8,260	7,654	61,024	79.99		
Shortfall		-4,737	-28,975	-38		

Source: Tabulated using data from Power System Statistics 2015-16

Cost of generation is higher than the allowed tariff, compelling the government to pay subsidies, which it promises but does not pay regularly and totally. Resultantly, IPPs do not make fuel payments and hence the term circular debt. When the PML-N government came to power in 2013, it did pay off the accumulated circular debt, but more was created as the basic disease of higher cost of supply compared to tariff is still there. Perhaps this may never go, as essentially the cost differential is almost equal to the theft and receivables. If both poor and rich steal and manipulate the political system, there may not be scope for improvement until the social conditions change.

Transparency

But why should a political government near the elections not pay and worsen load shedding. It is not understandable. To find real and deeper answers, one needs to have reliable data that is available to all and here comes the relevance of transparency.

We have National Power Control Centre (NPCC) that has all the kind of relevant data. The first thing a clever water and power secretary does is that he orders NPCC not to share data even among government organizations.

They never shared the NPCC data with the Planning Commission despite efforts on the part of the latter and shared probably only concocted figures. The data should be available on websites for all so that real reasons and culprits are identified. In most industrialized countries, this data is publicly available, sometimes free and sometimes on paid subscription.

People would keep groping for answers and would not get close to facts and actual situation. Even the government is susceptible to deception and fraud. When the politician discovers the inefficiencies and mis-statements, it is too late. They by the time have promoted the wrong guys and removing them late has no consequence.

The explanation of the water and power minister is also correct that due to seasonal misalignment of snow melting and hot weather, the hydroelectric power production is much lesser than the usual. The million-dollar question that needs to be answered is that what amount of load shedding is due to mismanagement and inefficiency i.e. what amount of load shedding could have been avoided with better efficiency and without alleged policy of controlling circular debt and subsidies and thus producing less. If the latter is true, I think there would be nothing more foolish than this. I don't think, they are that foolish.

On transparency, I would urge the prime minister and the water and power minister to order online publishing of NPCC data daily so that he is not fed with concocted figures compiled manually and credibility of government data and statements improves for the benefit of all.

32. DISCO Performance

NEPRA has released Performance Report for the year 2014-15 recently which gives us an opportunity to have some discussion on the issue. Following broad conclusions can be made as a result of examining the report:

1) None of the DISCO is up to the mark in terms of performance criteria (which we will discuss later in this space).
2) No company has improved its overall performance (except minor improvements) over the last five years; in fact, there are cases of worsening performance
3) In relative terms, IESCO appears to be the top performer among two close followers, GEPCO and MEPCO. Average performers are LESCO (quite a questionable ranking as we shall discuss later) and FESCO. Below average performers are K-Electric, PESCO and HESCO; and the worst ones are QESCO and SEPCO
4) In provincial terms, as is the foregone conclusion, Punjab located companies are better performers of the lot.

NEPRA report is based on the following criteria: a) T&D losses; b) recovery of dues; c) time frame for new connections; and 4) safety. NEPRA has given equal weightage to all these four issues. One could arguably differ with equal weighting of, say, time frame for new connections vs. T&D losses, although it is quite difficult but not impossible to assign a differential weightage .The simplest and easiest task with the authors of the report possibly was to give equal weightage to the most important and the least important criteria.

Reliable data is available only for T&D losses and the recovery as these relate to accounting. For example, T&D losses merit a weightage of 50 as against equal weightage of 16.6 for the remaining three criteria. Data on other criteria is rather questionable. The rationale for different weightage is enhanced due to data quality issues as well. The report has acknowledged the data quality issue and has excluded data on interruptions (SAIDI and SAIFI).

With this preface, let us now discuss in some detail, the companies' performance.

T&D Losses: The worst performers in terms of T&D losses are PESCO and SEPCO. SEPCO has reported a loss figure of 38.29 % as against allowed losses of 27.5 %. Similar is the case of PESCO with a reported loss of 34.8% as against allowance of 26%. NEPRA allows tariff adjustable losses based on the difficulties and objective conditions prevailing in various companies.

There is a negligible improvement in terms of these parameters in both of these companies. The improvement shown may be the margin of error in measurement itself. The irony is that the governments in Sindh and KPK raise hue and cry against MoWP and the DISCOs if and when punitive action is taken against powerful defaulters and thieves. The only recourse available to take action against unidentifiable defaulters is to disconnect all the customers drawing electricity from a distribution transformer. K-Electric uses this technique which rewards and punishes areas in terms of load-shedding and quality of service; DHA gets better service than Lyari or Liaqatabad. Reportedly, there used to be a time when large scale theft was in these areas.

K-Electric and HESCO, although among worst performers, have reduced their T&D losses considerably; K-Electric from 32.2% to 23.69 % and HESCO from 33.8% to 27.1 % .T &D losses of QESCO, MEPCO and LESCO have increased. QESCO loss increase may be understood in the circumstances. However, there is little justification of LESCO and MEPCO for their deteriorating performance in this respect; MEPCO's losses have increased from 15% to 16.7% and of LESCO from 13.3 % to 14.1% .In almost all of these companies, the reported losses are actually under reported, as the word goes by, and there are rumors of overcharging through fast meters. LESCO's deteriorating performance has a proportionally large impact on total country level losses as its share in total sales almost amounts to 20 % of the total. People of Karachi will have convulsions on this as they think that they are the largest in everything.

Figure 16 DISCOs T&D Losses (%)

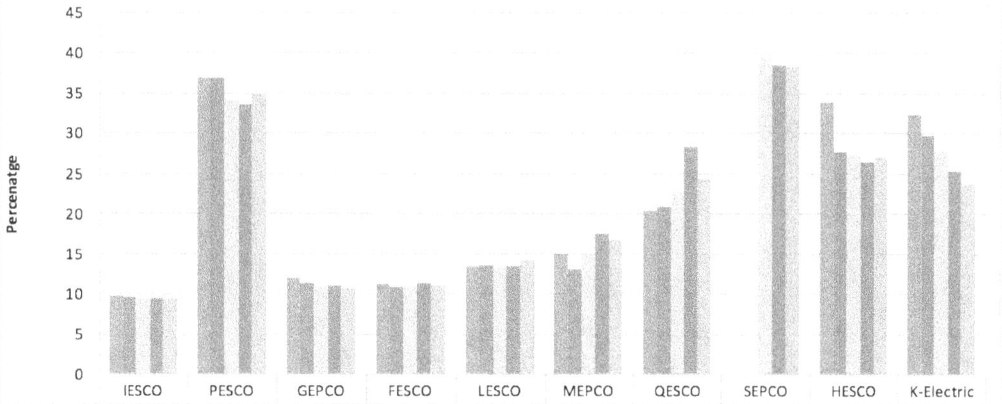

	IESCO	PESCO	GEPCO	FESCO	LESCO	MEPCO	QESCO	SEPCO	HESCO	K-Electric
2010-11	9.7	36.9	11.97	11.2	13.3	15	20.4		33.8	32.2
2011-12	9.52	36.9	11.23	10.8	13.5	13	20.9		27.7	29.73
2012-13	9.4	34.2	10.75	10.8	13.2	14.8	22.7	39.51	27.3	27.82
2013-14	9.46	33.5	10.97	11.3	13.4	17.5	28.3	38.56	26.46	25.3
2014-15	9.41	34.8	10.72	11	14.1	16.7	24.4	38.29	27.1	23.69

Source: NEPRA State of Industry Reports

Recovery of dues: NEPRA's State of Industry Report 2015 states that Rs 120 million have been added to the stock of recoveries taking the total to Rs 600 million. Recovery of monthly bills and arrears is also an important issue on which the prospects of success are better than the T&D loss reduction; dues and recoveries are recorded and identifiable while theft is difficult to pin-point and eliminate. Consequently, most companies have improved performance in this respect except QESCO and LESCO.

QESCO's recovery has slipped from 41% to 32.6 %. Perhaps only the poor pay in QESCO area. We may condone this due to the special circumstances prevailing in that province. But what justifies deterioration in LESCO, from a recovery of 95.88 % coming down from 98.1 % in 2010-11.They need smart meters on which we will return a bit later.

The nearby Gujranwala has been infected regionally and its recovery rate has also come down from 98.8 % to 97 %. There might be data discrepancy issues as well as there may have been under or over reporting in earlier years. The numbers of later year are plausibly more accurate as considerable effort has been made at both companies level and as well as MoWP level in improving accounting in this respect.

IESCO, FESCO and MEPCO have practically no receivables as per data of 2014-15. All of these companies have improved performance in this respect, although I have reasons to suspect the numbers of MEPCO. K-Electric's performance has stagnated over the years in this respect, as its recovery rate has remained static at around 90%. I do sincerely hope that NEPRA organizes special audits in respect of T&D losses and recoveries. There are many underlying motivations in these areas that merit rigorous scrutiny on the part of NEPRA and others.

Figure 17 Recovery Rate of DISCOs (%)

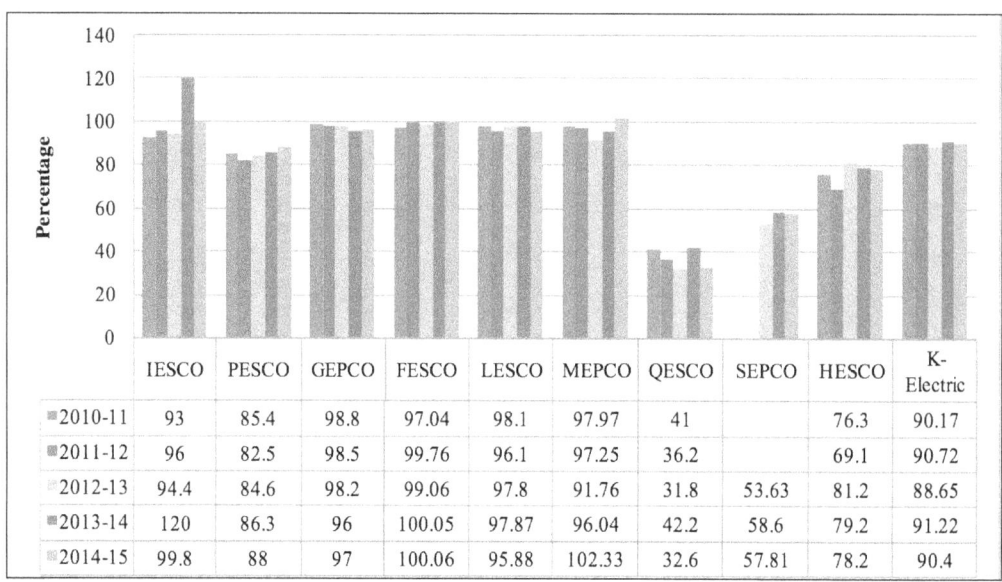

	IESCO	PESCO	GEPCO	FESCO	LESCO	MEPCO	QESCO	SEPCO	HESCO	K-Electric
2010-11	93	85.4	98.8	97.04	98.1	97.97	41		76.3	90.17
2011-12	96	82.5	98.5	99.76	96.1	97.25	36.2		69.1	90.72
2012-13	94.4	84.6	98.2	99.06	97.8	91.76	31.8	53.63	81.2	88.65
2013-14	120	86.3	96	100.05	97.87	96.04	42.2	58.6	79.2	91.22
2014-15	99.8	88	97	100.06	95.88	102.33	32.6	57.81	78.2	90.4

Source: NEPRA

It may be pertinent here to a case study from India where an India privatized DISCO has managed to improve its performance very significantly. In India, there are similar problems of load-shedding and theft and socio-political problems of poverty and power. Average T&D losses in India are at comparable scale with Pakistan at 30%. In Pakistan, privatization has not done much improvement in the performance of K-Electric. However, in India, it has made a difference. Let us take the example of Mumbai which is comparable to Karachi, in almost every sense, and where Reliance Group took over as a result of privatization. They have reduced the T&D losses to just 10 %. There are no UPS, voltage stabilizers, inverters or DG sets in MUMBAI DISCOM as claimed by their brochure. Customer satisfaction, as per independent agency estimates, has gone up from mere 22 % to 84 % in a matter of 7 years. SAIFI (average number of supply interruptions per customer per year) has improved from 9.65 to 4 interruptions per customer per year and SAIDI from 372.8 to 183 minutes per customer per year. For comparison purposes, SAIFI values for North American Utilities are around 1.1

interruptions per customer per year on the average. SAIDI (average interruptions duration per customer per year to be unreliable) values are 90 minutes.

Technical Performance: NEPRA has rejected the numbers given by DISCOs on SAIDI, although claims made by some of the DISCOs are not tall claims by any standard and as such these seem to be plausible. For example, for PESCO, SAIDI is 27,934 minutes, for MEPCO 15,677, for QESCO 7,507. These minutes do not possibly include interruptions due to load shedding. For FESCO, LESCO and K-Electric, SAIDI numbers are respectively comparable: 2,683, 3, 010, and 1,330. IESCO claims of 0.95 and GEPCO claims of 13.2 appear to be spurious (being better than North American Utilities) or they seem to have used some different method of calculation. Similar data issues are there in case of SAIFI.

Figure 18 System Average Interruption Duration Index (SAIDI)

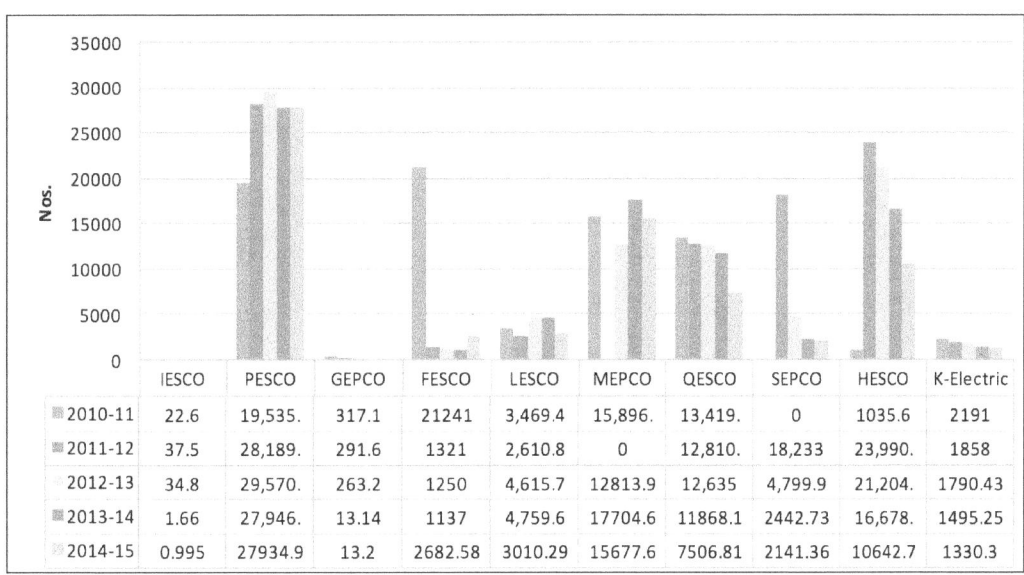

	IESCO	PESCO	GEPCO	FESCO	LESCO	MEPCO	QESCO	SEPCO	HESCO	K-Electric
2010-11	22.6	19,535.	317.1	21241	3,469.4	15,896.	13,419.	0	1035.6	2191
2011-12	37.5	28,189.	291.6	1321	2,610.8	0	12,810.	18,233	23,990.	1858
2012-13	34.8	29,570.	263.2	1250	4,615.7	12813.9	12,635	4,799.9	21,204.	1790.43
2013-14	1.66	27,946.	13.14	1137	4,759.6	17704.6	11868.1	2442.73	16,678.	1495.25
2014-15	0.995	27934.9	13.2	2682.58	3010.29	15677.6	7506.81	2141.36	10642.7	1330.3

Source: NEPRA

It is very vital that some seriousness is brought into these figures. Manual recording is often unreliable and there is a vested interest in not showing real numbers. Extension of SCADA up to 11 kV feeders and installation of monitoring equipment and smart meters at substations could provide the required data apart from improving actual performance.

It might be relevant here to discuss the smart meter project being financed by ADB (Asian Development Bank) under which 1.17 million smart meters would be installed in LESCO covering 35 % customer base and 800,000 smart meters in IESCO proving the same coverage percentage. Total investment in this project is of RS 490 billion. It would require no less than 5-7 billion USD and a period of 10 years to cover all the DISCOs. Can we wait that long for performance to improve and will we have that kind of finances?

Figure 19 System Average Interruption Frequency (SAIFI)

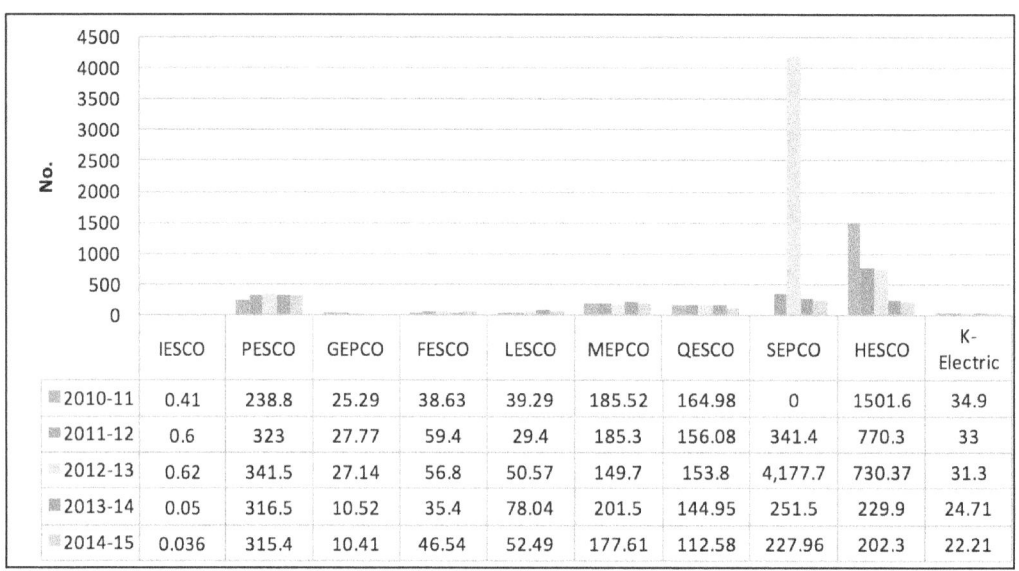

Source: NEPRA

An alternative approach would have been to cover only substations or even distribution transformers under the scheme and cover all the DISCOs. It would have cost possibly less than the present project. This would have not only identified DTs and Substations where the sales are less than the dispatch; but would have also provided much required automation and helped in improving O&M performance.

Planning Commission, succeeded in achieving some cost rationalization but its recommendations for modifying the project on these lines were not accepted by the stakeholders. However, another project can be launched on the lines suggested. It would not be expensive and can possibly be financed under normal PSDP budget. NEPRA may like to look into it, although under the threat of new legislation already, it may not like to interfere.

Let me suggest a very cost effective way of collecting data relevant to SAIDI and SAIFI and may be other performance indices with or without or very little involvement of DISCOs. It can be done by involving market research companies to collect SAIDI and SAIFI required data from electricity consumers. An app can be developed for smart phones, which would record and transmit interruptions and their durations. 1000-2000 volunteers in each DISCO can be inducted with some incentives to collect and transmit this data as the event occurs. Market research companies routinely collect consumer data and provide reliable results. Things are changing. Media companies are being rated by advertisers for releasing advertisements to be shown before or after the TV programmes. Minds have to change, otherwise instruments and technology are available. In fact, annual consumer satisfaction surveys could be ordered by NEPRA to be undertaken by market research companies. In other countries, companies themselves organize such surveys irrespective of regulatory requirements.

Concluding, a lot of improvement is required in NEPRA report itself and in the data collected for compiling these reports. Indeed, it may be useful for NEPRA to engage independent consultants to prepare annual performance reports, which may be able to identify and highlights issues and problems, the resolution of which may go a long way towards improving DISCO performance.

There are many areas that are not covered by NEPRA requirements, for example analysis of complaints and their nature. Consumers normally complain of error in billing, difficulties in lodging complaints, locations of complaint centers etc. Analysis of such complaints and trend of improvement in this respect is a must for performance improvement. As mentioned earlier, Customer Satisfaction Surveys, data collection apps, and annual performance reports compiled by NEPRA or independent consultants can help improve performance. Finally, all of this is data. Somebody has to read it and take corrective action. Successive levels of management hierarchy starting from supervisors and line managers to senior management in companies and finally at the ministry level have to do what they have to do.

Is there a scope for improvement? Of course there is. Efficiency, Quality, Service Level and Profitability should be the yardstick of measuring company performance. In terms of loss reduction, IESCO, FESCO and GEPCO have reached levels, where further reduction may not come as easily as these companies are close to the target of around 8-10 %, although all other companies have a long way to go. With T&D loss reduction, company profitability would improve, as at this moment most companies are paying for a portion of T&D losses from their own pocket i.e. consumer tariff does not cover these losses. With increase in profitability, cash flow would improve and companies would be able to invest out of their own funds for expansion and improvement. NEPRA Report does not deal with company profitability and I have not seen relevant reports, so I would leave it at that.

NEPRA Report deals with service level, but there is failure in getting the right data. On cost side, important criteria would be O&M cost per unit, no of employees per kWh or per 100 customers, or per km of network or employment cost as a percent of sales. DISCOs have to improve and NEPRA itself has to improve.

Most of the worst performing DISCOs are spread over large geographical areas, which makes effective day-to-day management difficult and many vices prevail in far-off places with impunity. It may be worthwhile examining the proposal of making DISCOs smaller. Some partial steps have already been taken in this respect like carving out SEPCO from HESCO. There is a case of dividing PESCO and MEPCO into three companies each. IESCO may be divided into two companies, the new company in the area of Rawalpindi and downwards. In Gawadar, a new DISCO can be made for Western or South-Western Balochistan.

A more pertinent issue is whether DISCOs (as well as NTDC) would be able to distribute another 8,000-10,000 MW that will come on line in 2017-18. Both NTDC and DISCOs are already overloaded and cannot deliver the already available generation capacity. Almost 50% of all DISCO's transformers are over-loaded (loaded more than 80 %, causing tripping and energy losses) and 29% of 11 kV feeders are overloaded (in KPK, the situation is more serious, where more than 50% 11 kV feeders are overloaded, as per NEPRA's State of Industry Report 2015.

Power dispatches from Uch-2, Engro Combined Cycle, Altern Energy, Habibullah Coastal, Foundation Power and Nishat Chunian have suffered from power dispatch problems. More recently, many Wind Power plants ready to be commissioned cannot be connected to the grid. Relevant authorities keep saying that distribution and transmission capacity expansion projects are being implemented on a continuing basis and that as more capacity comes in, they would be able to handle it.

We have discussed some important proposals that may go a long way towards improving DISCO performance in the next chapter. Reducing the company size should also make a difference.

Table 49 Salient Parameters of Distribution Companies

	Consumers	Peak Demand	Sales	Losses	Area
		MW	GWh	%	Sq. Km
PESCO	2,956,567	2798	7,599	34.81	77,474
TESCO	441,562	560	1,101	21.68	27,220
IESCO	2,462,167	2347	8,147	9.41	23,160
GEPCO	2,923,493	2386	7,055	10.72	17,207
LESCO	3,909,862	5021	16,328	14.1	19,064
FESCO	3,445,357	3091	10,806	11.03	36,122
MEPCO	5,116,072	2892	11,711	15.5	105,505
HESCO	976,888	1167	4,020	27.08	81,087
SEPCO	722,392	1357	2,682	38.29	56,300
QESCO	564,887	1762	3,994	23.1	334,616
Total PEPCO	23,519,247		73,443		
K-Electric	2,158,290	3056	12,293	26.89	6,500
Pakistan	25,677,537		85,736		

Source: NEPRA

33. Reorganization of the Power Sector

There are two opposite but probably correct characterisations that have been made regarding the performance of the power sector and the parent ministry (Water and Power). A large number of generation projects have been successfully launched and about 10,000 MW of electricity would be available by 2017-18. No denying that it is a good performance although, it is more of a political achievement rather than administrative. Nevertheless, effort is required to implement such fast track plans.

It is however, alleged that the short-term issue of load shedding was not managed properly and that load-shedding could have been less under a more efficient structure and leadership. IPPs remained under-utilized and not to talk of GENCOs, for which the ministry may pass on the blame to the economy managers who put an arbitrary limit on subsidy and circular debt and did not make the expensive RFO available. Despite all kind of irresponsible discussion by the TV anchors and apparent openness, there is still lack of transparency and dissemination of correct data and information, based on which objective analysis can be made. Hence, we cannot be the arbiter of this controversy, as to who is responsible. It appears highly unlikely that economy managers would be naïve enough to ignore energy, while on another plane they are allocating a lot of debt resources to the same sector. Blame has to be shared, probably, by both. Or it is the undue centralisation and accumulation of power in a few hands that is responsible. In this article, we will focus on the disintegration of the power sector, which in our opinion, has played a major role in the less than optimal performance of the sector.

Power sector has experienced disintegration under the utopia of the IFIs (international Financial Institutions). Their idea was essentially based on reducing the role of government and bureaucracy by bringing in private sector. However, privatization in Pakistan has been limited. If the recent statement of the new secretary Water and Power is correctly reported by the press, privatisation plan has been deferred or shelved indefinitely.

Before restructuring of the power sector, there used to be an intermediary level of professional management, intervening and communicating between government and the operating companies. Now bureaucracy in Islamabad is directly managing the companies and handling the day-to-day affairs. Ministry should instead only be concerned with policy and ensuring that rules and procedures are followed.

The Utopia of independent Boards

Another local and perhaps foreign utopia was that companies Board of Directors would be strengthened and that company autonomy would lead to improvement in performance and efficiency. This has not happened either. How can part-time board members, meeting occasionally, be expected to have the kind of impact that was expected of them? Senior board members are represented by junior proxies often. Board members often are reported to be liabilities on the Chief Executives, asking all kinds of favours and privileges and thus foregoing any leverage they might be having otherwise. Often the Chairman of the Board, who is Secretary or the Minister speaks and controls the board. No useful discussion takes place, except autocratic approvals. There may be rare exceptions, but this is a general rule and trend.

What is the result? Privatisation has not happened and independent and professional board of directors could not be established. Companies are being directly managed by the bureaucracy sitting in Islamabad interfering in day-to-day operations of the companies and ordering transfers and removals. Secretary Water and Power has emerged as a czar of the power sector. This kind of direct power and its accumulation and centralisation has created many problems of performance and efficiency. Secretary, being a human being is bogged down, sometimes due to personal choice and tendencies and sometimes due to shear load and circumstances.

Misconception about the role of the regulator

Another misconception was about the role of the regulator. It was wrongly expected of the regulator to control the performance of the companies. A regulator is an adjudicator among the competing and counter-acting interests of the consumer and producers. How can it, for example, enforce the merit order despatch when gas is arbitrarily allocated to less efficient plants by the bureaucracy? It can issue strictures and levy fines but cannot manage, control and coordinate the companies which is to be done by the owner or its representative institutions or bodies. On the other hand, regulator is disliked, if not despised, by the bureaucracy and even ministers have spoken against it and attempts have been made to prune their wings. Although, regulator has a quasi-judicial function, it should not be treated in the same isolation as normal judiciary. Neither, regulator should feel shy in communication and interaction, nor should the bureaucracy consider it as an adversary. Regulator should appreciate the difficulties of government and respect its policies, while government should seek advice and input of the regulator as the regulator has a lot of knowledge about the structure and economics of the sector.

Who says that individual companies can perform better if left alone? There are large utility companies like EDF, Iberdola, Enel, and Engie, EoN etc. managing and coordinating among hundreds of companies, which together employ hundreds of thousands of employees and generate revenues of billions of dollars. There are large corporate offices manned by professionals and experts pooling technological and human resources. I stand for privatisation where it can work and deliver and where it is politically and otherwise possible. What has been achieved by the announced privatisation of PIA and Pakistan Steel, HMC and other institutions? Privatisation is announced but not implemented and the organisation is left to decay to nothingness.

As it has happened in the case of power sector, centralisation and bureaucratisation increases by the removal of intermediary profession at management institutions like PEPCO. I have never been associated with PEPCO and do not think that it had ideal performance. But you have to have professionals managing the sector and not the general bureaucracy ordering around in highly technical institutions. There are two models: one that the ministries create a technical human resource base and structure within the ministry and the other is of creating an intermediate corporate set-up. The first model has been adopted by MPNR wherein Directorate Generals have been created manned by technical persons. The other, corporate model in the form of PEPCO was there in MoWP which was dismantled resulting in bureaucracy directly managing.

It is high time that the disintegration of the sector is done away with and pooling institutions are created for managing power companies which can coordinate better and pool know-how and resources. There are common problems, which have to be seen in its essentials and totality and there are common resources spread broadly which have to be put together to solve those common problems. I am not a conspiracy theorist or a believer of such theories. But there are many in this country who think that the foreign advice was meant to increase our dependence and impede our industrialisation. It may not be intentional. But the job of thinking cannot be subcontracted to advisors and consultants of foreign origin who neither have the commitment and nor the appreciation of the local problems. Fortunately, due to the grape-wine of CPEC, the role of IFIs would be reduced and there would be more freedom in doing one's own thinking and making one's own choices. Otherwise in good old days, a few hundred million dollars' loan used to be accompanied by a long list of mandatory advice of the kind I have lamented in this piece.

Merger of AEDB into PPIB

Good news for the power sector is the merger of AEDB into PPIB which was long overdue, especially, in the context of 18th Amendment, wherein provincial governments have been empowered to initiate and approve renewable projects. There was some kind of Khichri of institutions and a tri-partite approval process had to be put in place to satisfy the triplicity of the institutions. V.S. Naipaul writing on South Asia, said: recognising a problem, South Asians make a department, install a sign-board and think that problem has been solved or will be solved. Perhaps, creation of AEDB was slightly more than that. AEDB manpower and expertise would continue to be there, it is not being dissolved, and in fact would benefit from the larger resource base of PPIB. It would eliminate duplication or triplication and investors would have to deal with one less organisation, meaning a reduction of 33% of their time and resources. I am not supporting this only because I have been arguing about it in my publications and in board meetings of AEDB and counselled the Minister of Water and Power in this respect. They keep quite usually in response, and later implement such proposals quietly. Similar was the case of NEPRA policy changes wherein my policy draft proposal was largely adopted. I must clarify that subservience of NEPRA to MoWP bureaucracy was not included in my proposal. I should be happy even thankful that my proposals have been finally accepted, instead of complaining. Any way, it is good to see GOP bringing about changes. It is hoped they would consider the proposals made in this section as well.

34. Federalism in Electricity

Electricity is largely a federal subject in our constitution with some role for the provinces in small-scale power generation and the sector being under the oversight of Council of Common Interest (CCI), although some confusion persists as evidenced by the formal and informal role of provinces in policy making and investments. Sindh Chief Minister, Murad Ali Shah has threatened or warned HEPCO, SEPCO and WAPDA to pack up or stop the disconnection operation in Sindh. He added that Sindh Government can handle its electricity sector itself. It is not the first time that a chief minister of a province has expressed the desire for autonomy in the electricity and energy sector as a whole. Similar statements have been issued earlier by KPK on issues related to tariff, hydro royalty, autonomy etc. Although self-control and management is a longer-term concern, more practical issues are at stake momentarily. Small provinces are a beneficiary of pooled pricing as we shall see later, but the provincial governments are behaving as free riders by not supporting the elimination and control of electricity theft drives launched by DISCOs.

We are a federation like the U.S., Canada, Australia and Germany, with a difference that the U.S. has a Presidential system, while all others including India and Pakistan are Parliamentary systems. Our erst-while Colonial master, the U.K., has a unitary system, something that we tend to forget while trying to ape our masters. In developed countries the confusion and controversy of federalism and self-control are much less limited due to the role of the market. In federations, electricity generation, for example, is out of government control whether federal or provincial. It is market based with prices being determined by specialized commodity exchanges. Transmission and inter-state trade is normally under federal control while provinces can have provincial transmission under their control. Distribution is almost everywhere under provincial or even local government control. However, federal or provincial control in market economies does not mean ownership and management also. Companies normally own and operate the facilities but the regulation is with government only.

The problem becomes much more complicated in countries like ours including India, where most of the electricity sector is owned and operated by governments and when it is the government, there is this controversy of federal or provincial. In Pakistan, we have cooperative federalism. We do not leave the provinces on their own, but share the burden of each other. For example, there is one price of petrol or of electricity from Karachi to Chitral. In other commodities, such as fertilizer and wheat, perhaps, there is a somewhat similar system. In power sector, price of electricity, which comes from various sources of generation, is pooled and averaged. This has a lot of merit and has served us well, although WTO types might disagree.

Pooling has its problems as well, as we are seeing in the case of electricity sector, which is beset with the problems of losses such as theft and receivables. In Punjab (except MEPCO whose losses are 20%), the losses are generally around 10%, while in all other provinces the losses are in the bracket of 25-35 %.I a way, Punjab subsidies other provinces with respect to losses. However, things are more complicated than this.

Hydro electricity is cheaper in KPK, while gas based electricity is cheaper in Sindh and Balochistan. KPK nationalists argue that hydroelectricity of Rs 3.00 (earlier of Rs. 1) is taken from them (although not exactly, as Tarbela and others are all federal investments) and resold at Rs 10. But if you ask them, as to what they would do in winters, when there is no hydro, they have no answer. KPK royalty has been doubled as now it gets around Rs 1.00 per unit. They are flushed with royalty income, so much so that they are investing the royalty income in arguably inefficient projects.

It is shifting sands. Commodity based nationalism or sub-nationalism has short life or rationale. Bangladesh was made, in large measure, over conflict on jute. In 1971, jute earned a lot of foreign exchange. Today, jute has been superseded by plastic. Earlier, Balochistan used to be a major producer of gas which is no more the case as Sindh is the largest producer today which may again be superseded by RLNG and fast upcoming of KPK gas fields. Ultimately, renewable energy such as solar may be replacing, oil, gas and coal. Minerals used to be a major concern of political economy and raw material for conspiracy theorists. Recycling and synthetics have diluted the role of minerals and will continue to do so.

Power sector is a capital-intensive sector. A lot of external political issues influence this sector. It is highly unlikely that provinces alone would have the wherewithal including finance, management and other resources to go alone in this sector. Look at Thar. Despite Sindh province's successful struggle of gaining full control (which has wasted a lot of time by the way), Thar coal and power projects have taken up so much time. Federal guarantees had to be given despite a 49% private equity and a higher tariff Moreover, all Thar projects would come under CPEC, and otherwise there would be no other avenue for exploring Thar coal.

Federal government would have to restrain Punjab government from insisting on installing imported coal power plants in Punjab. If not, where would Thar electricity go? Where is the market and who would invest in Thar? So, there is a merit in cooperative federalism of sharing strength and weaknesses and opportunities and threats. Specially, when all of us are week and poor, we need cooperation more. I do hope that we grow so rich that one brother may not need the other, as is the case usually.

Federal government has offered in the past that Sindh government takes control of electricity distribution and even Pakistan Steel Mills. Interestingly the wise men of Sindh, only want to have management and not bear the cost and want to continue to be a free-rider .I have currently, no reason to defend the federal government but truth has to be told .It would be childish to try to take over DISCOs and Pakistan Steel, just for the sake of it and afford losses to the tune of billions.

Smaller provinces should stop acting as free riders and cooperate with the federal government in reducing losses and theft. All are involved, poor and rich and private sector and government departments. These losses cannot be reduced by active cooperation of the provincial governments. In smaller provinces (and to some extent in large one as well), it is often difficult for the provincial governments to take action against the powerful, both rich and poor. Thus wiser approach would be to let the DISCOs continue with their campaigns and indulge in lip service to support the locals. I hope Mr. Murad Ali Shah is doing the same. In Sindh Government, there is nobody who is able to understand these issues, as he does, having the right experience and qualifications.

We are passing through an energy crisis. Hopefully, by 2018 this should be over with the completion of new projects. After this period is over, some tinkering may be possible with the system. In the long run, all provincial and federal governments may consider and evaluate alternative models such as that in India, where there is larger provincial role in the electricity sector and do a cost-benefit analysis. Electrical sector in India, under provincial domain partly, is not an ideal condition. In Indian model, generation and, within province, transmission and distribution are managed by provincial boards. There is a provincial regulator and independent provincial tariff and thus no pooling with other provinces. There is however, inter-provincial trade option.

There is load-shedding all over, despite having coal and other advantages. We should try to improve on the current system by improving the operations of CCI (Council of Common Interest), while federal government should take steps to reduce the informal and de-facto influence of the Government of Punjab on its determinations.

MPNR has commissioned a restructuring study of the Petroleum sector. MoWP may consider commissioning a similar study, which may involve consultations with the provinces. Neither decentralization should mean anti-federation nor is the talk of centralization to be taken as patriotic. These are administrative issues of which cost-benefit analysis and consideration are to be more important than any other consideration in order to maintain a sector that is viable and is able to attract investments. CPEC has temporarily solved the problem. This grapevine may not always be there with that kind of funds. Indeed, CPEC's energy sector investment has been possible, among other factors, due to the availability of a reasonable and well-structured power policy that has been developed and improved upon over the years. One has to remain open and eligible from all sources.

Excessive centralization and preponderant role of Punjab thereof led to the passage of 18th amendment. There are people who are not satisfied with the implementation of the 18th amendment and would like to expand its scope while there are others who have doubts whether the 18th amendment has been successful in delivering to the provinces what was expected of it. It is true that there are capacity issues in the provinces. A review appears to be in order with a view to improve upon it. As it is said that problems of democracy can be solved by more democracy. In the same way the proponents of autonomy and decentralization argue that the failures cannot be made an alibi to wind up or dilute the process of decentralization .It is the competitive aspect of decentralization that is often forgotten and has to be stated here.

Under a federal monopoly, there is no competition. Under decentralization, provinces compete with each other, not for federal funds, but for investors. The latter are often at the mercy of one federal entity, while in provincial systems they have more options and flexibility. Secondly, governance has a lot of bearing with decentralization. Bureaucratic structure has not been altered accordingly in the wake of 18th amendment. Provincial bureaucracy is largely limited to grade 20, while provinces are now required to handle the whole sectors of Education, Health and others independently of the federal government. It cannot be done by junior officials. There is a case for upgrading the provincial bureaucracy to the federal level of grade 22, at-least in selected sectors.

Planning Commission has recently prepared a report on Governance Reforms. I am not sure if some consideration has been given to this aspect. Forces of status quo appear to be prevailing these days for whom classical system is untouchable, if the recent decision on the appointment of DCs in Punjab is an indicator.

Provincial role in the Energy sector

Under 18th amendments, provincial role in the energy sector has been recognised. Some powers have been given to the province like project approval of small renewable projects, and representation in Directorate General of Oil and Gas. Provincial roles are further enhanced by the role of CCI (Council of Common Interest) which is empowered to take all the major decisions.

The situation is confused and complicated. Government of Sindh complains that it is not given its due role and its projects and recommendations are not accepted or approved. KPK has been lamenting about its hydro royalty issues which has since been most or less resolved. Provinces (Sindh and KPK) want more role in new projects approval and want to have some independence in this respect. They complain that GoPb has been given undue preferential role. GoPb has launched its own RLNG projects and made GOP to establish two (and now three with recent approval) RLNG plants in Punjab. They also complain that GoPb has been pressurising Federal government to accept more and more imported coal based projects in Punjab to the detriment of Thar coal. Solar power projects have been launched without prior formal approval and tariff obtained after implementation. It is also complained that bank of Punjab, a GoPb entity makes undue profit through lending at high rates to the energy projects which is to be paid by all.

However, the counter argument that there is no formal preferential power granted to GoPb. It is due to a practical situation that Chief Minister Punjab and the PM come from the same party and the family and that due to the Energy Crisis, informal but practical role has been given to the former for chasing and expediting the projects. The result is fast track implementation of Sahiwal Coal Power Plant and the fast progress in all the three RLNG projects.

However, all politics is about power and all politics is local, as is generally accepted. Politicians want power as it helps grab more power. In case of energy projects, powerful investors, promoters and landowners make all kind of proposals and want to have special power to steal electricity and withhold bill payments inordinately (another type of pseudo-legal theft).Many sceptics and cynics argue that power brings money as well, but it may also be to enable oneself to serve the electorate genuinely. It can be both, to be fair.

Recently, there was an issue of gas and electricity transmission of a Sindh government project. Sindh government has launched its own transmission company. There are issues of integration and power purchase agreements. There are powerful landowners and promoters having interest in Wind Power projects which are stalled due to tariff issues and impending reverse auction of renewable projects. An appeal of Sindh government with respect to exemption of wind power from competitive bidding and reverse auction has recently been rejected by NEPRA. And very recently, Sindh Chief Minister lobbied rather heavily for getting bagasse power projects approved and even offered to buy the electricity (of-course at high tariff) and pay by deduction at source. He has been misguided by the powerful landowners and sugar mill interests. The Bagasse projects would take two years to complete, in which time frame, already a surplus is being forecast and thus the reluctance of the CPPA to enter into a liability. Who benefits from it? GoS has major issues in social sector. Why is it interested in assuming additional liabilities which the federal government is absorbing, willy-nilly? GoS is not alone. KPK government is diverting its hydro-royalty income to making investments in questionable hydro projects. KPK is full of abject poverty. It badly needs investment in social sector and direct attack on poverty.

Sindh government has been mulling to launch its own electricity regulator. But then who buys and pays. There are a lot of subsidies that go to the electrical sector. Electricity losses and theft are the highest or one of the highest in Sindh. Would Sindh be able to have its own Power Purchaser and pay for the losses of DISCOs in its territory? Why should it want to do that, when Federal government is carrying this load? On the other hand, why should the Federal government keep carrying the subsidies load and transfer the sector to provinces? One would lose financially by accepting liability and the other gain by doing away with liability? Some solution and modus operandi has to be found to satisfy and reconcile the competing political interests. After all it is a federation. Self-control and autonomy is priceless but it has to be balanced against the requirements of development and growth of all in the federation. There are issues of scale economy and capacity.

There are four kinds of federalism, which are modulated by the increasing role of the private sector: centralised devolved, dual, and interactive and dynamic federalism. In a market economy, role of government becomes minimal, but the monopoly nature of utilities still requires federal government involvement in policies and even in control. We moved from a centralised model to a devolved or dual model under 18th amendment. Most countries are practically moving to an interactive and dynamic model of federation where there is some overlap and redundancy in order to be able to deal with the various political, social and political issues and challenges. Germany is a special case in point, whereby through party processes, interactive and dynamic federation has been implemented. Ironically, GoPb's activism can be interpreted in the same spirit?

The Alternative Provincial Regime

In most federations e.g., USA, India, Germany, Australia and Canada, electricity sector is largely provincial and local. Role of private sector has led to much lower role of governments; the latter is mostly regulatory, advisory and perspective planning. The regulatory role is more pronounced in inter-provincial exchanges, pricing and delivery issues or are there in safety issues of nuclear and hydropower and dams. Federal and provincial regulators co-exist. There are strong electricity regulators at state level in India, the U.S., Canada and Australia. There is no common power purchaser, however, which in our cases buys expensive and sells cheaper and assumes liability. Resultantly, there is a wide difference in electricity prices in all these countries; in India and the U.S., this is highly pronounced. The question is that would it be in the interest of all including provinces, to have differential electricity pricing. We have a very deeply entrenched uniform pricing system which is based on subsidies and transport charge sharing etc.

Electricity payment and subsidies issues will get pronounced with doubling of the electricity availability and sales, although power supply quota issues may go away with surplus. Everybody gets what it ever wanted or will want. However, new issues of pricing would emerge, e.g., who gets the cheaper hydro energy, which by the way would not be cheaper in near future due to rising royalty impact and increasing construction cost. Solar is abundantly available, in all provinces which augments solar's competitive position as a resource. Federal government may examine ways and means of how it can unload the subsidies to its beneficiaries which, however, would dilute its very source of power. Initially, one could consider, passing on half of the subsidies to provincial governments, as half of the subsidy or more goes towards compensating the electricity theft. Presently, provinces have no interest in curtailing and combating electricity theft. In fact, it used to be itself stealing and let both poor and rich to steal. Poor have votes and rich have power.

India, initially, made State Electricity Boards (SEB), controlling the affairs of India, initially, made State Electricity Boards (SEB), controlling all the affairs of electricity in the individual states. They have moved towards converting boards to autonomous corporations to promote commercial and business like operations. There are government owned transmission and distribution companies that function under SEB(C)s. Generation can be both government owned and as well as private IPPs. PPAs are signed between DISCOMs and IPPs directly and bilaterally. SEBs funnel the subsidy to DISCOM, if any. There is a state regulator dealing with most of the Tariff issues. As mentioned earlier, federal regulator CERC (Central Electric Regulation Corporation) deals with residual and inter-provincial matters etc., including inter-state transmission.

In case the alternative provincial model ala-India is adopted, it would become even more essential to have a central institution for coordination to counter fragmentation and pool-up the human and technical resources. India made Central Electric Authority CEA) for this purpose. There will be capacity issues in smaller provinces which may like to remain under federal cover. The cost and benefits of such a fragmentation has to be weighed. Apparently, federal government would be relieved of subsidies budgeting and provinces would get the power of self-control which some of them are craving for and demanding so vehemently. P.M. Nawaz Sharif had offered such a conversion reportedly orally. There was no response from provinces. They also wanted control of PSM without liabilities and opposed privatisation. Authority without responsibility is perhaps the most enjoyable thing in the world.

Role of Energy/Electricity Market

Perhaps provincialization has to wait till an electricity market is established in Pakistan. In Electricity market regime electricity is bought and sold in a share market kind of bidding. Companies can buy and sell electricity and as well as provinces can. IPPs can be launched without permission, except for safety and technical issues. There can be merchant IPPs who sell to every customer at the prevailing market price. Pakistan has committed to IFIs under reform programme to bring about Electricity market. An experimental market regime can be set up by allowing IPPs to sell 10% of their production to market. On a similar model, two electricity exchanges are operating in India. A natural gas and LPG market can also be set up on similar lines. Wheeling regime both in case of electricity and gas has already been set up, although there are still issues therein.

Concluding, GOP should bring in an institutional normalcy in the energy affairs and undue influence of individuals and provinces in policy making and important decision to the exclusion of other stakeholders should be avoided. Energy crisis required some crash processes which should not continue indefinitely. This will greatly reduce provincial sense of discrimination that may result in hasty and unplanned fragmentation.

The need for coordination in the Energy sector

There is a further need of integration and coordination in the energy sector as a whole. There were utopias of merging Ministry of Water and Power and the Ministry of Petroleum into one Ministry of Energy and the merger of NEPRA and OGRA. I myself have been guilty of making these suggestions. The requirements of integration and coordination can be met without physical merger of the organisations and ministries. There is an Energy Committee of the Cabinet in which the energy ministries and Planning Commission are represented. The committee has played a vital role in managing CPEC related issues but could not be effective in handling load-shedding issue. For this coordination at working level is required among various energy institutions. There was a proposal to organise Federal Energy Forum in which all energy institutions could be represented, which would meet monthly for exchanging information and making (possibly non-binding) recommendations. Discussions and flow of information may hopefully result in automatic adjustments at the required levels, while orders may be defied, resisted or ignored.

And lastly, while reconstruction and integration of the sector is being recommended by building intermediate management institutions, it is equally important to divide and distribute large DISCOs like MEPCO and PESCO into smaller companies so as to reduce their geographical coverage and thus be able to handle their business more efficiently.

It is an irony that while aiming for privatisation, we have ended up in the obverse. We are vacillating between the extremes. Reconstruction of the power sector has become vital. May be it is too late now, as elections are approaching. It should be the priority of the next government, for many challenges still remain such as theft and losses which would increase in absolute terms with increase in generation capacity.

Alternative Provincial Electric Governance Regime

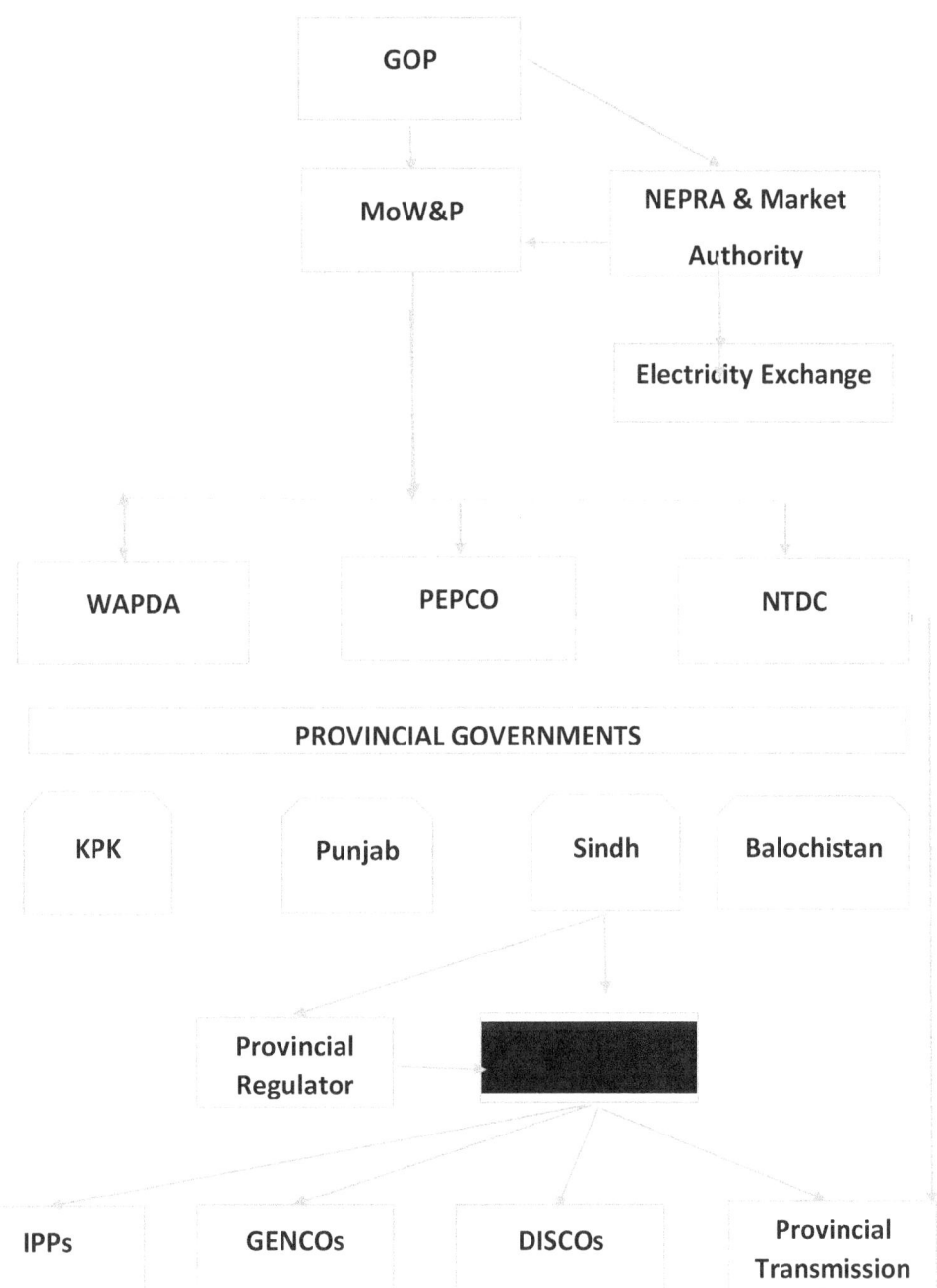

35. Reviving PEPCO

Many changes are going on in the energy sector; recently, the two ministries of power and petroleum have been merged into a unified ministry; now directives have been issued for subdividing DISCOs; earlier power sector reforms were introduced diluting the role of regulator. It appears that this round of changes are indigenous and home-cooked and have not come out directly from financing agencies or their consultants. We will in this space venture to make a case for reviving PEPCO.

About two weeks ago, P.M. Abbasi who is also holding the charge of the Ministry of Energy (MOE), issued directions to divide PESCO and carve out a new DISCO for the Hazara division. The directive missed out MEPCO with possibly the largest geographical coverage. Now Minister Laghari (of Power division) has issued a comprehensive order that includes MEPCO and other DISCOs. The implementational aspects, such as of boundaries, are to be decided by the respective boards of directors.

In the earlier round of reforms, WAPDA's role was considerably reduced to include only hydro power generation and water issues, and PEPCO was totally dismembered. WAPDA has been further pruned down to deal with hydro power only, as a new Ministry of Water has been created, which may be a step in the right direction, as the subject of water is too important to have a divided attention of one ministry. It deserved a unifying one ministry dealing with all issues of water exclusively, which has been achieved through the creation of water ministry.

The proposal of dividing the DISCOs, and for that matter gas companies also, into smaller companies in terms of organisation and geographical domains, has been on the table for a long time now. Most of the vices, it is said in the two sectors of power and gas, came from the lower tiers of these organisations and it has been argued that smaller setups may give more direct and closer controls to the top management.

Carving out a separate DISCO for Hazara division has attracted opposition from the relevant circles alleging political undercurrents, as there have been initiatives and demand for a separate province of Hazara earlier and there were controversies and debate over naming of what is now KPK province which used to be NWFP. Technical arguments have been given for dividing on load basis rather than geographical basis, although the division logic emanates from geographical contiguity. May be additional geographical entities can be added to Hazara division to do away with the underlying political suspicions.

A common theme of reform proposals for power companies and even other public sector companies has been to make the board of directors independent, powerful and competent enough to undertake supervisory oversight functions or even manage the strategic affairs. This has remained largely a utopia .PEPCO was dissolved under this utopia. Neither independent and competent board members could be inducted due to the political and social system prevailing in the country, nor has the bureaucracy been able to provide a conceptual and operational framework of the operations of the board. Senior bureaucrats often are not able to attend the board meetings and study various company proposals, and participate in special committees. Boards' discussions and meetings are dominated by the chairmen and are mostly a perfunctory exercise attempting to meet the legal requirements. Practically, the companies are being managed by the bureaucracy from Islamabad. This is true more for power division entities like DISCOs and GENCOs. Whereby individuals may have a lesser inclinations for direct interventions, the CEOs alone function, without a meaningful oversight functions from the board of directors. In fact there is a vacuum and confusion as to who should be controlling DISCOs; ministry or NEPRA, as confided by Chairman NEPRA to me. NEPRA, as per its own understanding of its functions, it is responsible for broad oversight and perhaps rightly so. Actual controls rest with the board of directors and the ministry. However, Ministry thinks that NEPRA is responsible for the bad performance and lack of initiatives by DISCOs and the former is responsible for administrative and transfer functions.

PEPCO used to function as an intermediary between Islamabad ministry and the DISCOs. They used to have a CEO and technical department which could understand the technical and operational issues and could provide guidance to the entities, examine their reports and have an oversight over them. Rebellious and independence minded CEOs with the benefit of the utopia that we have mentioned earlier managed to sabotage PEPCO and got it dissolved in collusion with the interventionist bureaucracy of Islamabad. Under the current system, there is what the poet said; Na Khuda hi mila, Na visale sanam.

Over the years, power sector has been restructured and dismembered. It has gone from one extreme of a monolith WAPDA to another extreme of smaller splitted organisations which are not able to accumulate expertise and develop and absorb technologies and skills. Resultantly, there is hardly any major initiative or proposal which does not emanate from foreign consultants and experts tendered by IFIs (International Financial Institutions).We need an accumulator organisation. This can be done by reviving PEPCO which should control DISCOS and GENCOs as well. Although some GENCOs (burning Furnace Oil) will be closed down, many new DISCOs would emerge. It would become well-nigh impossible to control and manage all these organisations from the ministry in Islamabad. Large companies controlling many organisations have similar central organisations even in the U.S., Europe and Japan. In India and Korea, there are integrating organisations of this nature. By reviving PEPCO in a slightly altered form as proposed earlier, the current extreme would be balanced.

36. Regulatory Reforms

Two actions have been taken by the Ministry of Water and Power, of late causing quite some stir among the stakeholders. First, NEPRA has been put under direct administrative relationship if not control of MoWP, in place of the Cabinet division, which may not be as consequential as is being perceived by the public. Second, a summary has been moved to CCI (Council of Common Interest) to reduce the powers of NEPRA and bring about some associated changes. While the first step of administrative reporting appears to be catering more to psychological needs of the power ministry bosses, the second step is of more serious concerns and consequences.

Proposed Changes

We would like to take a stock of the proposed changes in a non-partisan manner and undertake an objective and cool analysis. Some of the proposals are meant to remove barriers to entry and initiation of the market process, and bringing an element of planning into the system, which perhaps everyone would support. The other proposals are more or less aimed at reducing the role and power of NEPRA, which may be contentious requiring more consultation and expert input for bringing some balance. Following amendments have been proposed:

1. Doing away with Generation licensing requirement
2. Introduction of Market Operator and Electricity trading
3. Development of a National Electricity Plan and Tariff Policy
4. Appellate Tribunals to hear appeals against NEPRA determinations and orders
5. Curtailment of NEPRA jurisdiction on certain aspects of transmission and distribution companies
6. Authority to determine Tariff components and not the total tariff
7. Doing away with provincial representation in NEPRA Board
8. Ministerial guidance or directive to be binding on NEPRA

In this chapter, we shall be analyzing each of the proposed changes for its merits and demerits.

Doing away with Generation Licensing

The amendment proposes complete doing away with the generation licensing irrespective of the size of the facility. Electricity Act 2003 of India also has similar provisions. They continue to have this provision in the law. Had they been facing difficulty or adverse consequences, they might have repealed the provision. Many people argued that this is a time waster and a barrier to entry. For large-scale power generation, licensing may be desirable and a necessity, for it may have bearing on reliability, safety and other public interest issues.

However, for smaller power plants, say under 5 MW, whether in utility sector or in captive generation, licensing may be done away with, as it would eliminate unnecessary expenditure of time and resources. Unfortunately, it may cause some reduction in the revenue of the licensing authorities.

Complete de-licensing of rural electricity generation and distribution is also called for. There are many waste-to-electricity producers, micro-hydel power plant owners and others, who are operating under a legal vacuum and are victim of blackmailing of utility staff and other government agencies. SMEs operating in electrical sector may not be equipped technically and financially, to bear the load of licensing process, fee and paper work. This step should be taken in this perspective and not in the context of curtailing power of the regulatory agencies.

Introduction of Market operator and Electricity Trading

This marks the advent of wholesale open market activity as opposed to regulation where tariff and prices are determined and fixed by the regulator like NEPRA and OGRA. Electricity markets normally operate like a stock exchange or any commodity exchange where spot prices are determined through matching buyer and seller offerings. This appears to be a good beginning, although there are many a slips between the cup and the lip.

Electricity or energy markets are evolving in many developing countries, while in the developed countries, it is the norm. There is no regulation on generation in these countries. India also has two or more electric exchanges where electricity is bought and sold for the next day delivery. Perhaps 10 % of electricity generated has to be sold by law on these exchanges.

Eventually, in Pakistan similar provisions would have to be made. Initially, regulators may also be given the additional function of controlling the Market Operators. Eventually, when the role of open market increases, a Market Commission may have to be made and NEPRA may be a part of that commission, representing or dealing with a smaller sector. However, Hanooz Dilli door hast.

Electricity Policy, Plan and Tariff Policy

This is a new addition to the NEPRA Act. The existing Power Policy is in fact a Tariff Policy. Electricity Policy may be developed separately. The main issue is the Electricity Plan. NTDC used to make a projection that may be akin to a plan. Currently, there is no official Electricity Plan that may be binding on the project approving institutions. The plan would provide a common tool to PPIB, NEPRA, CPPA, NTDC, distribution companies and the government. In that such a provision should be welcomed. Strangely, NEPRA's role in the preparation of this plan has been provided for. Perhaps it is a compensation for removal of NEPRA's nontariff jurisdiction on transmission and distribution companies. Thus, this is an attempt to formalize decision making which has remained more or less informal and arbitrary.

Electricity Plan would be for five years and may be amended. In a way, this is a reinstitution of the erstwhile Five Year Plans, which became out of fashion for more than a decade now. It won't be easy to make a reasonable plan and still more difficult to follow. Planning Commission has tried to do it in the past under ADB's assistance, but failed. They have revived the project under a different framework. JICA under MoWP project has developed some projections. Earlier Canadian consultants used to advice NTDC on it. Varying Projections have always been there in the system. This time, there is a provision in the Act itself, which may bring some formalism into the process.

Appellate Tribunals

The amendment proposes a full-fledged tribunal with a Chairman and two members who would hear appeal against NEPRA's determinations and orders. The tribunal would be authorized to modify or annul NEPRA orders. This may be taken to be a big affront by the regulatory body, but such tribunals are not unusual.

Currently, there is a provision of review process. If a party is aggrieved by NEPRA (regulator), it can file a review petition with the regulator, and by drawing the attention of the regulator to new facts and figures, may try to get a modified decision from the regulator.

What is the point of appellate tribunals then? There are several reasons in favour of the tribunal. First Appellate tribunals are common in other sectors such as in case of tax, service matters in Pakistan and in many other countries. Review petitions in current practice are heard and adjudicated by the same persons in NEPRA due to the paucity of human resource.

Appellate Tribunal offers an opportunity of noninvolved persons and experts to examine the contested cases and announce their judgment. To avoid unnecessary costs, there can be one tribunal for both electricity and Petroleum sector including coal. This is not unique. Appellate Tribunals are working quite satisfactorily in India for quite some time now.

Curtailment of NEPRA jurisdiction on certain aspects of transmission and distribution companies

This is the most debatable and contentious issue and has drawn criticism shadowing the good parts of the Act. NEPRA powers have been curtailed in the following ways:

1. The word "Prescribe" has been substituted by "Specify". That is wherever NEPRA used to prescribe, it would specify. Lawyers would be better able to explain the difference. Prescribing appears to have more powerful implications than specify. Deliberate effort have been made by drafters for this substitution. There may be some objectives behind it or could be a simple effort at belittling NEPRA.

2. Transmission and Distribution activities have been removed from NEPRA domain in non-tariff terms. Under existing Act, NEPRA has a lot of say and jurisdiction on T&D companies such as preparation and submission of investment plans and others. Since, T&D companies are government owned unlike generation IPPs, which are mostly private, this appears to be an attempt to get an unhindered and exclusive control on the T&D companies. This amendment may be useful only till the time privatization of T&D companies does not take place. Once privatization takes place, who would compel private companies to undertake serious investment planning and action. Thus it appears that privatization is not under active consideration of the drafters or has not been thought through in an enthusiasm to regain exclusive control.

Authority to determine Tariff components and not the total tariff

This may be considered the most important and crucial and perhaps divisive and contentious one. There is a public hue and cry on putting NEPRA under direct administrative control of MoWP, which may only be vexing personally to the management of NEPRA. However, real curtailment is embedded in the tariff domain. Under the proposed amendment, NEPRA would determine Tariff components only and not the total tariff. Does it mean, that government would determine the total tariff by adding subsidy or surcharges? There are pros and cons. The provision would instill arbitrariness in the system and may reduce investor confidence. The advantage is that government may be able to include policy preferences and objectives in the tariff system. More thought and debate should be organized to finalize on this under a committee as proposed by NEPRA.

Doing away with provincial representation in NEPRA Board

The professional and expertise requirements of the Members are very demanding. I am not sure how the dual requirement of merit and provincial representation are met. Do provincial governments organize a transparent process to select a nominee? There is no such process as I know about it. I am sure merit is compromised. And then, how do such representatives respond to dual loyalty? I am not sure if provinces have benefitted from such a provision except favoritism opportunities. The proposal of doing away with the provincial representation of one each from every province is a step in right direction. However, quota requirement could be included, one member each from provinces, to maintain a sense of provincial participation.

Issue of Guidance issued by the controlling Ministry

Ministries are responsible for formulating policies while regulators and others operate within that framework. Sometimes this distinction is blurred. Government makes the policy in energy sector and is finally responsible for all good or bad .It has to provide the resources and has to face the public. Sometimes Government may feel that the regulator is impinging on its role or the regulators decisions are creating problems in the government functions and sometimes reverse may be true. There are many models. Some models give a high role to the regulator and some give more role to externally developed government policies.

In the current regime, MoWP has reason to feel that its DISCOs are being controlled by the regulator to the extent that it tantamount to management. MoWP intends to build some safeguards in this respect to improve efficient working of companies owned and managed by it.

Why is an Independent Regulator Necessary?

Regulatory function is a quasi-judicial function. Investors, consumers and stakeholders should be confident that their interests and rights would be fairly protected and a balance would be maintained amongst opposing requirements. If investors suspect that the government or other powerful parties may pressurize the regulator to get unfair treatment for them undermining the others, they would feel higher risk and would either refuse to come in or would require higher returns compatible with this regulatory risk, which would be of no benefit.

Regulator is supreme in an autonomously working system unaided by government subsidies and support and where many private sector buyers and sellers deal with each other. Naturally, that requires a powerful and independent empire .In the current system, there is a lot of government support and subsidies, and government controlled actors among buyers and sellers. Naturally government would like to have more say in regulatory domain, as it is the government which has to face the brunt and consequences of the regulatory action. Unfortunately, there is no formal mechanism for consultation or adjudication between government and the regulator, which created difficulties in the relationship between the ministry and NEPRA. Public hearing is the only consultation mechanism, which is unilaterally controlled by NEPRA. In other jurisdictions such as India, there is a Regulatory Advisory Committee, which provides a mechanism of consultation among government, buyers and sellers, consumers and the regulator. Unfortunately, even the proposed amendments do not include RACs. Appellate Tribunals would play some consultation and adjudication role, hopefully.

This perhaps is the most delicate and controversial subject. While Ministerial Directive politely put as guidance may sometimes be necessary on policy issues, it can compromise the independent functioning and image of the regulator. It may also open the ways for corruption and nepotism etc. Let us see what OECD guidelines on regulators say on this subject:

Where legislation empowers the Minister to direct an independent regulator, the limits of the powers to direct the regulator should be clearly set out. The legislation should be clear about what can be directed and when. Any directions made by the ministers or politicians should be documented and published (OECD: Principles for the governance of regulators, 2013).

Thus Ministries guidance may not be a taboo, but the guidance provision may not to be a cart-blanche for interference on a daily basis and limitations should be built into such provisions .Spelling out such limitations may not be easy, and would require consultations under expert guidance .OECD assistance can be obtained through the good offices of USAID or others.

Regulatory Advisory Committee

This is my own proposal meant to improve the quality of consultation process. Regulators (NEPRA/OGRA) usually hold public hearings on all its determinations, whether for award of tariff or Licenses. There is a structured process of interveners, commentators and general participants. However, often it has been observed that such meetings are not adequately attended by all stakeholders. Mostly, industry representatives are there .If public is there, the participants are not experts or specialists to be able to respond and participate protecting the interest of the group they represent. To solve this issue, many countries including India have provision in their requisite laws for organizing   Regulatory Advisory Committees on which stakeholders and experts are represented. At least board and lodge of RAC members are paid by the government or the regulatory commission itself. RAC is provided a space in the regulatory Authority premises to hold its deliberations. Thus the determination process benefits from a wider resource and consultative process. The current tussle and difficulty that we observe between Ministries and Regulators is the lack of third parties and institutions such as RAC and the Appellate Tribunal. This is my own proposal and is not part of CCI summary. In various forms and styles, such RACs are currently operating in Canada and India. There may be other countries as well following this practice.

NEPRA Act was legislated in 2003. Nothing remains static while the context and circumstances keep changing. Thus amendments may not be a taboo. However, such amendments need to be discussed in an open and transparent process. Government moves are not as abrupt as these appear to be. Planning Commission proposed reforms in the power sector about two years ago and circulated a draft among the stakeholders including the provinces. Quite a few of the amendments appear to have been drawn from that draft.

NEPRA has demanded due consultation and expert input, which should be heeded to by the concerned quarters. The proposal of making a committee with a deadline is a reasonable one. Sky is not falling requiring any bulldozing. On the other hand, politicians should not make it an issue for point scoring. They should, however, insist on due consultation.

37. Alternative Approaches to Smart Metering

Smart meters are new generation of electric and gas meters having a memory chip and an electronic communication device integrated with a communication network like GSM, GPRS, and Broadband , Wi-Fi etc. They can do many things: send hourly consumption data to utilities, monitor demand and connected devices, change allowed loads and disconnect and reconnect from the computers centrally installed in utilities. They eliminate need of errand meter readers and enables disconnection of defaulting consumers, and reconnecting once default is removed, without the need to send a lineman who is often found to be corrupt aiding and abetting in electricity theft. Just as Income Tax reforms have eliminated direct contact of taxmen with the taxpayers considerably reducing corruption and nuisance, smart meters may prevent corruption by eliminating the need of utility employees running around the lanes and doing all sort of bungling. DISCOs would be able to monitor theft at higher professional level and take appropriate measures.

One approach is to install smart meters all over in a matter of 3-5 years and start benefiting from the afore-mentioned advantages. There are around 35 million consumers in Pakistan, and in traditional high capital-intensive approach, one meter has an installation cost of about 25,000 Rs per meter. For a 90% coverage, this would mean 787 Billion Rs (7.87 Billion USD) of investment in total or 1,577 million USD per annum. Can we afford that? At this moment we have many other pressing needs like improving transmission and distribution infrastructure, even if we assume that generation investment would be financed under CPEC.

Under the traditional system, a project has been launched financed by Asian Developed Bank, under which about 1.5 million each smart meters would be installed in two DISCOs covering about one-third of the consumers each. There were serious policy issues and concerns of the stakeholders which were only partly resolved and the project was approved and is now ready for implementation for the last six months or more. Even if this fast track program involving more than 500 million USD is implemented tomorrow, where would the DISCOs be? Would they be anywhere near to preventing theft and improving performance? One would not be too sure of it, as only one-third of the consumers would be covered under the program. Would another so-called tranche of ADB be readily available, and when would the august time, of at least 90% coverage, will arrive? Until then, other DISCOs would remain uncovered, where the problem of theft is much more grave and severe. What should be done or should have been done? It is not easy to come up with answers and solutions.

The answer lies in a composite of alternative approaches consisting of; a) adoption of cheaper and affordable technologies and approaches; b) starting with DISCOs where the theft issue is more severe like in MEPCO, PESCO, HESCO and SEPCO, even if we don't touch QESCO in Balochistan; c) Starting in all afore-mentioned DISCOs simultaneously with a less capital intensive approach in order to have some palpable impact within 5 years, the tenure of one political government. Quite confusing and complicated? Let me explain, how?

First of all, all developing countries have not caught on the bandwagon of smart meters, at-least in the kind of approach that has been adopted here in Pakistan. Small development programs have been launched to judge the right approaches. In our case, even though small programs were launched under a USAID program in MEPCO and PESCO, these remain isolated. The new program designers have not bothered to look into the positives and negatives of these programs.

A lot is happening in the field of smart meters and smart grid. New communication technologies, open systems and new software approach like cloud systems are chipping in. Developing countries are particularly examining the possibilities of retrofitting an add-on module of the size of a cigarette packet on top of the existing static electronic meters. It hardly costs one-third of the smart meters and is installed conveniently. It has been noted, examining the cost structure of the smart meter projects that overheads are three times the cost of the meters itself. Dismantling the existing ones and installing a new system plus the contractor approach are responsible for the overheads. In this case we have found dis-economy of scale rather than the usually found economy of scale. Perhaps, a retrofit solution implemented and installed by DISCO staff gradually may be cheaper than a fast track contractor based approach. A similar approach has been adopted by K-Electric, although without retrofitting. Retrofitting technology has matured over the past few years and has been adopted by India and Brazil and many more are in the process to do so.

Then there is the gradual approach, involving first installing smart meters up to the Distribution transformer and subsequently adding smart grid features. In my view, it may be worth examining if 90% coverage of DTs would provide a better handle on identification of theft. If most DTs have a smart meter, one can identify the areas where dispatch is more than sales indicating theft. Or a mix of the two approaches, covering, say one-third consumers and 90-100% DTs or as the budget may allow. If one summarizes the afore-mentioned, it is as follows:

- Open system
- Standardization and combined procurement by DISCOs
- Retrofit solutions
- Covering DTs first and adding smart grid features
- DISCO implemented, non –contractor, gradual approach,
- financed by DISCO from internal resources
- Possible subcontracting of Billing

One may find the proposal of subcontracting the billing function contradictory with the earlier discouraging and non-involvement of contractors. The rationale is that IT competency is generally not available in government controlled DISCOs, while installation skills and efficiency are. Many utilities in Europe and the U.S. have adopted subcontracting of billing. Also the driver of a gradual approach is based on the need of protecting our local industry. In very large contract packaging that is proposed in the current project, the chances are that the local vendors might not be even short-listed due to the prequalification requirements of large internationally funded projects. Their only chance to participate would be as smaller parts of a joint venture wherein cream goes to the more powerful. When the locals are the leaders, chances of technology development and absorption are much higher. At some point in time, we will have to start thinking in these terms. India and Malaysia have done this with the obvious results.

Retrofitting by self-implementation appear to be the most desirable and essential step. Although, one would not like to put a wrench in a going project, those responsible may reconsider their approach and examine the plausibility of the above submissions. One would like that NEPRA may also take some interest in it and add diversity to the leadership of the project.

38. CASA - a liability?

Independent economists have been cautioning against mounting debt and have issued forecasts of balance-of-payment difficulties and of other associated issues. Recognizing that investments have to be made to solve the problems such as energy deficit, the least that can be done for tackling these issues is to invest in projects, which have good pay-offs. GOP and particularly MoWP have done a good job of bringing so much power into reality, as is indicated by advance partial commissioning of the two coal power plants. They deserve all the praise. While, the supply situation is clearly under improvement, it is high time to review some of the projects, which are not optimal and were conceived in altogether different and difficult circumstances. A realistic and objective evaluation of the project is intended here. The input and criticism in this respect is made in good faith and in national interest, which should be received with an open mind by the relevant quarters.

We will examine here CASA in the evolving power surplus position, security and safety risk due to continuing instability in Afghanistan, emerging electricity economics in the context of low oil prices at-least in the mid-term scenario, and possible negative foreign exchange impact among others and suggest measures to improve the project, and postpone the project if not cancel it altogether. In fact, there is a need of conducting a review of a number of other projects, as we shall see in this space, which have lost relevance in the changed circumstance.

Rationale for CASA

CASA is a project planned for importing electricity from Tajikistan and Kyrgyzstan through a 750 kV HVDC transmission line 1000 km long passing through some of the most difficult terrain. It will bring 1000-1,300 MW electricity into the system.

 CASA was conceived more than ten years ago at a time when oil prices were high, foreign investment was scarce and electricity deficit was on the horizon. It was kind of viable at oil prices of the order of 120 USD per barrel and more. It was a short-term solution to be implemented within two years may be by 2010. Following rationale was advanced to justify the project;

1. There is a surplus in Tajikistan-Kyrgyzstan area.
2. It was a cheap or cost effective solution.
3. It required no or very little investment or debt liability from Pakistan.
4. It would substantially reduce Pakistan's electricity deficit.
5. It would promote trade in the area and create mutual interdependence.
6. Eventually, law and order situation in Afghanistan would improve for peaceful operation of the project and that there would be no residual threat of physical attacks on the rather vulnerable transmission line. Afghanistan would be a beneficiary of the project thus government of Afghanistan would have vested interest in keeping the project safe from vandalism and terrorism.

Do we still need CASA?

None of the above assumptions are valid today. The project rationale was that there was a surplus in Tajikistan in summers, as is usually the case with hydro power. This surplus is, however temporary and would evaporate with time, in 10 or 12 years with the increase in demand there. The evidence is that as opposed to original 30 years supply arrangement as the PPA is now designed for 15 years. Where would Pakistan investment in converter stations go for the remaining 15 years life is an open question. World Bank most probably would advance loans to Tajikistan and Kyrgyzstan in order to install more capacity there so that they are able to export for the remaining period of CASA contract. In other words, the World Bank financing that would otherwise have come to Pakistan would be diverted to Tajikistan.

It is also no longer a cost effective solution. It might have been in the days of oil prices of 120-140 USD per barrel, but no more. CASA electricity would now cost as per latest agreement around Rs 12 per unit as opposed to Rs 6-8 per kWh typical cost of generation. Its PC-1 had been prepared and justified under basket price of Rs 15 per kWh plus which has come down to less than Rs. 10 per kWh. The project cannot possibly be approved by ECNEC under these circumstances having negative returns.

The assumption that it would require no or very little investment from Pakistan is also not true. An investment of some 300 million USD is required by Pakistan to install the converter and transformer station in Pakistan. This would be financed through a World Bank loan under CASA project. The truth is that all CASA investment is a liability of Pakistan, if not through debt finance then through a mandatory purchase PPA. The net effect is the same, an annual liability/outflow irrespective of electricity demand in the form of mandatory payments and dole money to Afghanistan.

That it would reduce deficit is not true anymore. With a 10,000 MW projects to be completed by 2018, and many more projects in pipeline coming thereafter, there is no justification for CASA. The sense of surplus is so much that MoWP has put a hold on signing of PPAs for Solar and Wind Projects. In effect, one is blocking solar and Wind power to facilitate CASA. Solar and Wind Industry and Sindh government would not like this policy. Even, Punjab government would like to install more solar. Either CASA would remain underutilized itself or will render other projects under-utilized, which would result in unnecessary payments. Already renowned economists are doubting our capacity to pay and forecasting bankruptcy and external financial crisis. Where is that hidden money (to justify such laxity and profligacy) that our MOWP bureaucracy thinks that we have and others are not able to see it.

The projection that it would promote trade and interdependence may be theoretically true. In a recent conference in Islamabad organized by Asian Development Bank, most participants/speakers said they would prefer energy self-sufficiency than trade in electricity. Indeed, the representative of Afghanistan was more emphatic about it. Government of Afghanistan has said, time and again, that instead of CASA, it would like to install a power plant based on its local coal. Electricity trade may be attractive and viable in the context of geographical contiguity. Afghanistan not being interested and enough demand not being there, transmitting electricity over 1,000 kms appears to be a simple lunacy.

Things have not improved in Afghanistan and nobody knows when it will improve. Successive Afghan governments have been against Pakistan including the current and the previous government. As for Taliban they have made their intentions known, when they attacked an existing electricity pylon passing through Baghlan province through which CASA would also pass through. Investment in a project that depends on Afghanistan for its survival is a big risk that one should not take. Apart from financial issues, interruption of CASA connected with Pakistan network, would damage Pakistan's supply system. In Balochistan, it is our own country and our utility staff usually repairs the damage quickly. Afghans cannot do the repairs themselves, and team from Pakistan may not be able to go to the site that swiftly and may not get effective protection as well.

A trouble free operation of this project in current or probable future is not likely. The recent round of terrorism originating from Afghan soil should be an eye opener for those who have been soft-peddling the security aspects of the project. In these circumstances, it would be very naïve and fool-hardy for the vested interest circles to still be in favour of this project. Reported involvement of PowerGrid of India in the project on behalf of Afghanistan or otherwise may even compound our risks. To improve terrorism survivability of the project, undergrounding of HVDC had been proposed by the Planning Commission. Apparently, this has been found to be expensive. Reportedly underground HVDC costs are coming down and the project could have waited until its economics improved along with the improvement in other conditions.

Indeed an alternative project design should have included installation of a coal power plant in Afghanistan of an appropriate capacity to be linked with CASA to improve its utilization and two way flow of energy within Afghanistan and beyond to Tajikistan , Kyrgyzstan etc., in winters where there is a power deficit in that season. That way Afghanistan might have become part of CASA. Actually networks are feasible for geographically contiguous and closely knitted regions where there are multiplicity of sources and sinks rendering high utilization of facilities. Project DESERTEC (bringing renewable energy from North Africa to Europe) has been shunned precisely due to similar reasons as are being argued against CASA.

CASA electric tariff is twice as expensive as other competing options. In the wake of the overall low price regime that came to emerge after the price finalization, electricity supply price should have been renegotiated which has not been done. In any case marginal costs should have been the basis of price negotiation than the average. Perhaps in the context of Rs 18.00 per unit of the cost of oil based electricity, such finer points were not relevant or required. Similarly, inflated payments have been agreed to be paid as transport royalties and doles to Afghanistan.

Most international energy agreements in the world have been renegotiated in the world that was made prior to oil price plunge. The same is required in this case including payments to Afghanistan. Competitively priced electricity is a life and death issue for our export industries. And then there are supply surplus issue raising questions about its utilization combined with poor security situation in Afghanistan render this project highly undesirable .We are a poor country facing many challenges. We are not in any position of indulging in largesse and charity .Our renowned economists are concerned regarding the debt repayment ability of our economy of the much needed projects, not to talk of such profligacies as CASA.

Concluding, there appears to be hardly any merit or justification of the project. If diplomatic sensitivities do not allow withdrawal from this project, it may be postponed till an appropriate project design and improved political conditions emerge permitting revival of this project based upon an improved project economics.

39. New K-Electric Tariff

NEPRA has recently announced new Tariff for K-Electric, which will be valid for the next seven years. The last tariff determination was made in 2009, which was almost a renewal of the tariff originally issued in 2005 immediately after its privatization. In this article, we would attempt to appraise the tariff determination in terms of its strength and weaknesses and make some recommendations for possible improvements. Following are the basic parameters of the new Tariff:

1. NEPRA has issued a base tariff of Rs 12.07 per kWh. K-Electric had asked for a tariff of Rs 16.23 as against the existing one of Rs.15.57. Effectively, there is a reduction of Rs.3.50 as against the existing tariff and a shortfall of Rs. 4.153 as against the demand of K-Electric. The approved tariff includes all the three components, namely generation, transmission and distribution. K-Electric is organized as an integrated utility which model is getting out of fashion due to market reforms that have been done in most market economies.

2. Loss target for the year 2022, is 12.53%, quite a good target, if it is finally achieved. This means that over a control period of next seven years, T&D losses would be reduced by 9.57%. This should bring K-ELECTRIC in a comparable position with other better performing DISCOs in Punjab.

3. Rate of Return (RoR) has been provided to be 13.27 %, which in the case of DISCOs is 15%.

4. Claw backs provision (sharing of profits with consumers), if company profitability crosses the threshold provided) is maintained. To be fair the tariff should be non-discriminatory and it should be fairly balanced against opposite risks. For example, if there is a claw back mechanism guarding against extra profit (which as indicated earlier may not be relevant in the present determination), there should be a protection against loss. For example, if loss targets are not achieved for some good reasons, the actual returns would not fall below a defined threshold.

5. Write-off of bad debts (receivables) has been allowed on actuals but limited to 1.83%; provisioning has not been allowed, although it would have promoted efficiency. Provisioning could have been limited to be adjusted against proof of write-off.

6. Efficiency improvements are as per Business Plan and tariff accordingly adjustable. There are no unknowns.

K-Electric had asked for continuation of the MYT-2009, with some additional proposals such as allowing returns on Revaluated Assets, which is extra-ordinary. It could have been allowed by NEPRA as a general incentive to all, and may do so in future, as an incentive measure to all. It could not have been permitted to K-Electric alone. I am not sure why K-Electric lobbied for continuation, for there have not been outstanding profits under the existing Tariff. Profitability is still much under the norm of 15-17%. May be it would have allowed them to recoup their efficiency investments, an issue that may remain contentious.

My assessment is that this tariff determination is no longer a performance based constant price tariff, as opposed to the MYT 2005 and 2009, which provided for a constant tariff of around Rs 4.50per kWh, (inflation such as in fuel price and others adjustable) while the benefits of loss reduction and thermal efficiency gains were to be pocketed by the company. It had no separate provision of RoRB, but was built in implicitly in the constant price tariff. Thus this tariff determination is cost-plus on the lines of existing DISCOs and GENCOs. Annual revenue requirements and investment plan have been projected by the petitioners, which NEPRA has scrutinized and accepted with adjustments as is usual with cost-plus. And three separate components for generation, transmission and distribution have been determined. The determination has been made on the basis of a business plan submitted by K-Electric. Thus everything is known. There is no uncertainty or risk except for the attainability of loss reduction as opposed to the MYT-2005/2009, in which many things could have gone wrong or right. Thus there was a provision for consumer protection against a windfall profit over and above certain norms.

So why should consumer representatives or K-Electric worry about claw-back mechanism and the rates and sharing formula? The RoRB (rate of Return on investment as cleared by the regulator) has been fixed at 13.27 %, which, it can be argued, is discriminatory. It should be comparable with DISCOs RoRB of 15% and based on 17% RoE for generation components ala IPPs. I have a different approach in this respect, as has been indicated in the afore-mentioned. I stand for a higher (2% more than the norm) RoR on loss reduction projects vis-à-vis routine or growth related investments. In the invest plan proposed by K-Electric, there will be a loss reduction investment of Rs 93 billion out of a total investment provision or undertaking of Rs. 287 billion. If higher return as proposed by me to loss reduction projects is accepted, the average RoRB may come out to be 15% compatible with other DISCOs.

I am not sure why K-Electric management is so fond of keeping the Generation component, which is a liability for K-Electric and a major irritant for public policy. Everybody is demanding for separation of the three components. Transmission can be clubbed with distribution, as it is a very small component. The residual issue is of generation. K-Electric can ask for provision in tariff for its existing assets and wriggle out of new generation responsibilities. K-Electric can always launch an independent IPP should it find it attractive as others are finding it attractive. I am not sure, if its argument of difficulties in cross- mortgaging assets in case of disintegrated operations is valid.

Overall, NEPRA determination appears to be all right, provided there are no issues in data and calculations. K-Electric does not appear to be sure of it and asked NEPRA for working details. K-Electric also thinks, that required investment may not be attracted under the instant determination. Consumer groups are not happy either, but it is not clear why? Both parties are preparing to request for a review. The axiom goes that when both parties cry, the arbitration or decision is correct and balanced.

40. Relending Charges in the Energy Sector

Prime Minister, Nawaz Sharif, has spoken recently about the need to reduce electricity prices. This was also a part of PML (N) manifesto. Although his claim of reducing electricity cost by 26% is correct, the reduction has not come about due to government action or policy but it is due mainly to the phenomenal reduction in international oil prices. It is not easy to reduce energy costs unless the subsidy is provided. Subsidies have in fact been withdrawn. Although there are not many opportunities for reducing energy costs, which are largely dependent on international factors, whatever little opportunities are available should be utilized to achieve this end. In this chapter, we will discuss some measures for reducing energy costs. We argue that by reducing the relending rates of the public sector energy sector projects of the public sector, energy costs can be brought down.

With the reduction in global oil prices, interest costs these days represent a significant proportion of energy prices in Pakistan. In pre-oil price collapse period, oil based electricity used to cost around 16 cents per unit with fixed cost (Capacity Purchase Price) being around 4 cents represented 25 % of the total cost. This ratio has today increased to 50%. Fixed cost essentially is the financial costs including interest and Return on Equity indicating the higher role of interest rates these days on the final energy costs and tariff.

Readers would be amazed, and certainly not amused, that the effective interest rates for public sector energy projects in the country are around 15-17% as opposed to international market rates of 4-5% and local currency rates of around 6%. The relending rates are the same even when local currency interest rates have come down from 14 % to 6%. The government borrows at the rate of 2-3% and relends to the energy projects at 15-17 % in return for services like guarantee and foreign exchange rate cover. The energy project pays in Rupees at the foreign exchange rate prevalent at the time of borrowing. Foreign currency in Pakistan, in the long run, has depreciated at about 5% per year. Thus government can charge this differential and guarantee charges (say 1-2%). Based on this, a legitimate relending rate would be around 8-10% as opposed to 15-17% being charged by the government. This has a direct bearing on the energy production costs and tariff.

Based on these high relending rates, gas transmission and distribution companies (SNGPL and SSGC) are paid 17 % as Return on Assets (ROA) in their annual revenue requirement, resulting in higher gas tariff. Neelum Jhelum Hydropower plant (NJHPP) capital cost increases are partly due to high relending rates. NJHPP costs have increased due to delays; inflation and interest during construction, wherein interest rates charged is 17%. Bhasha dam, RLNG projects, and other public sector hydel projects would also suffer from this irritant. In fact, the situation has become so acute, that public sector prefers to borrow these days from local banks instead, wherever feasible and some international loans have not been utilized because of this high-cost intermediation.

Planning Commission has spearheaded the demand for change in relending policy, which has been also supported by all the relevant public sector agencies and line ministries, perhaps, the only issue on which there is harmony or solidarity. However, if the case is so obvious, as I have explained, why do not the EAD and Ministry of Finance agree? Apart from ego (why it did not occur to me first, which I do not think may have come in between), it is the cost centered accounting approach that may be blocking change and reform. Under such an approach individual departments may have positive economics and earning, while the economy as a whole loses. Earlier, perhaps there was some rationale, if one is to be found, that subsidies were balancing the higher relending charges. Many argue that it is even more untenable, if not outright stupidity. Charging on one account and giving it away in other forms, while earning the criticism from the international financial institutions including IMF. And finally, it creates a differential or cleavage in the competitiveness of the public sector vs private sector, as the latter borrows internationally at 4-5%.

The issue is simple. I have taken all this space on elaboration for the sake of my readers who may not be specialists or may be lacking data or may not be aware of this anomaly. PM has spoken on reducing energy cost and tariff. It is hoped that the relevant bureaucracy and policy makers would pay attention to it and bring down the relending rates. This will certainly bring down energy costs and improve the image of public sector energy projects and will reduce irritants for new projects as well. Elections are approaching; this is the time for action.

Table 50 Pakistan Interest Rates March-May 2017

	%p.a.
SBP REPO	4.25-6.25
SBP Policy rate	5.75
Export Refinance	3
LTFF(end-user)	6
Treasury Bills cut-off (12m)	5.99
Treasury Bills WA yield (12m)	5.99
KIBOR(12m)	6.17
PIB (3 yrs.)	6.41
PIB (10 yrs.)	7.94
PIB-Secondary Market(10yrs)	8.09
RoR-RIC	6.54
RoR-DSC	7.54
RoR-SSC	6
RoR-PBA	9.36

Source: State Bank of Pakistan

41. Nuclear Power

Up till now, the general impression has been that the nuclear power is a competitive energy source, if not the cheapest, which has remained a contentious issue in the US Debate. Supporters of nuclear power manage to prove that it is competitive, if not the cheapest and opponents also manage to prove their case that it is expensive and uncompetitive. An additional factor that has come up recently is very high capital cost of the safer Gen-II nuclear power plants costing in excess of 5000 USD per KW. Pakistan is acquiring ACP-1000 nuclear power plants which are a variant of AP-1000 Gen-III reactors - Westinghouse type reactors. In this section, we will do a survey of similar projects in other countries with a view to examine the cost structure there and compare it with potential costs in Pakistan.

The proposed K2-K3 project would cost 9.50 billion USD with a Chinese low interest loan of USD 6.5 billion which appears to be the EPC cost as well. Admittedly nuclear power plants have become extremely expensive; 4000-5000 USD/KW vs 1600-2000/KW for conventional plants. Interest during construction in the case of nuclear power plants is usually very high due to the long time it takes (5-7 yrs) to put up a nuclear power plant. Actual financial cost would depend on the interest rates charged by Chinese, of which much has not been revealed. Chinese Finance companies have started behaving like any other western financing agency. They are requiring 4% or so of insurance charge on the lines of EXIM Bank of the US. Reportedly, on some projects (wind power) China has exempted Pakistan of this rather hefty charge. Usual net interest rates for such projects should be around 5%, especially as against a Libor of lower than 1%. Where would Pakistan bring the required equity of 3-3.5 billion USD from and at what rates (which appears to be the owners cost and interest during construction) is an open question?

However, nuclear fuel is cheap, costing 1.0 cent or slightly more per kWh as opposed to 16 cents for oil or 4 cents for gas. Also, due to a high capacity factor of 80- 90%, it gives more electricity per MW than other power plants e.g., twice that of hydel power plants and 30% more than other conventional plants.

In the adjoining tables, we provide CAPEX and COGE (Cost of Generating Electricity) data of nuclear power in a number of countries; a) Europe and the US; b) India and; c) Pakistan. The lowest COGE is in case of Indian-Russian design NPPS such as KundanKulam (India) commissioned recently with a price tag of 6.5 USc only, as opposed to 19.4 USc of the new Western supplied NPPs based on AP1000 technology of Westinghouse. DAE India is quite concerned over such costs although they have entered into advance implementation agreements. Most common estimates of COGE of NPPs in many countries hover around 12-15 USc. This data, however, belies the general impression that nuclear power is cheap. Capital Cost (CPP) alone is costing USc 10 or even more per kWh in DAE India is, however, rather desperately, trying to get arrangements under which nuclear electricity costs out of AP1000 power plants comes out to be 10-11 cents. The situation is even worse in the US California Energy Commission (CEC) predicts much higher COGE for Gen-III power plants.

- Merchant NPPs (IPP selling to open Markets) =USc 34.24/kWh (2018 nominal prices)

- Investor Owned Utilities-IOU NPPs(IPPs selling under long term arrangements)=USc 27.31/kWh

- Public owned utilities-POU NPPs=USc 16.68/kWh

Admittedly, there are methodological differences in costing. Gen-III reactors are being designed for 60 years as opposed to 30 years earlier. Table 1 gives NPP cost data from Western vendors that ranges around 5000 USD/kW. Table 51 gives cost data from China, Russia and India where, CAPEX is nearly half that of Western countries. Quality differences, financing rates and higher manpower cost may be responsible for such differences. One may note from Table 52, that China has been asking, reportedly, same prices as of Western sources of around 5000 USD/kW from Pakistan. There appears to be a scope for friendly negotiations as both sides do not have much choice. The Pak-China deal could be around the same terms as Indo-Russian deal for roughly the comparable technology, Russians having more experience. Russians have also provided 80-85% finance at 4% as well. If deal were structured around this cost data, one would have expected nuclear electricity from K2 reactors to be around 7 cents per unit, quite an affordable figure. In present case as the deal has been reported, it may be 70% higher, indeed quite uncompetitive and rather unaffordable.

A common theme would be apparent in the data on India and Pakistan. CAPEX both in India and Pakistan doubled every 10-12 years or with change in generation i.e. from generation I to II and now III. Up to generation-II reactors, nuclear power was competitive both in India and Pakistan i.e. gradually escalated to around 10 cents level. With the Gen-III reactors unit CAPEX has again doubled to 5-6000 USD/kW level bringing CPP (Capital Cost component) of COGE to be over 10 cents. With 1.25 cents for fuel cost and another 1.25 cents for O&M, the total COGE hovers around 15 cents. In order to

Table 51 Comparative Nuclear Power Cost from Western Vendors

Reactor Type	Project	CAPEX USD/kW	COGE Usc/kWh	Country	Origin
AREVA EPR	Jaitpur	4850	14.55	India	France
P1000	Mithi Verdi	6460	19.4	India	USA
EDF	Hinkley Point C		15	UK	France
P1000	Georgia Power	6360	10.4-11.5	USA	USA
P1000	Sanmen	3000		China	

Source: Compiled by the author, various sources including WANO

For 2018, following COGE forecast has been made:

address this situation, many buying countries are resorting to negotiations seeking discounts and other measures. India managed to get a 30% discount from Russia, while Turkey negotiated based on a long term tariff of 12.5 cents/kWh in its recent deal with Russia for its Akkuyu project. Unfortunately, our Chinese friends are charging the full US price, which nobody else may be able to offer in the international market. Pakistan can get nuclear reactors and the associated finance from China only for a variety of reasons, while due to rather lower credibility, no other country would buy nuclear reactors from China. Hence, sympathetic negotiations are required. China has installed similar reactor in Sanmen at a cost of USD 3000/kW, as we have indicated in the adjoining table. If prices of K2 and K3 are brought down to this level or slightly more, the nuclear power would become competitive. Otherwise the cost scenario would be inadequate. Pakistan has to bring down its cost of production to be able to solve its Circular Debt problem, as there is very little room to enhance power tariff anymore.

Table 52 Nuclear Power Cost in Pakistan

NPP	CAPEX (USD/kW)	COGE (USc/kWh)
Chashma I & II-C1, C2	2866	7.0 (PAEC)
Chashma III & IV-C3, C4	3000	9.59 (PAEC)
K2	5000 (PAEC)	9.59 (PAEC)

Source: Compiled by the author based on newspaper accounts, of Chairman PAEC statements.http://www.thenews.com.pk/Todays-News-2-224000-Pakistan-can-operate-larger-N-power-plants-PAEC-chief January 2, 2014.

Chashma I had a tariff of Rs 5.00 per unit (later enhanced to Rs 7.00) approved by NEPRA, while Chashma II is reportedly producing at close to Rs 10.0 per unit. Hydro is the cheapest (5 cents), followed by Natural Gas (Rs 5.00 at currently prevailing low gas prices, the situation would change drastically with the induction of LNG), Coal (7-8 cents). Nuclear Power would lie between Oil and others. Hydro and Thar Coal appear to be the most optimum choices for Pakistan in this scenario.

PAEC claims that K2's electricity would cost slightly lower than 10 rupees per unit (levellised-average). Initial costs may be twice as high during the period when financial repayments are being made. A new factor in nuclear energy economics is the long lives of NPPs extending up to 60 years. The real cost would depend on the financing rates; interest rates charged by China and other parties. If the interest rates remain under 5%, it would be possible to achieve the cost figures claimed by PAEC. Indians are worried about the fate of similar projects being supplied by the US and France to their country. It is patently clear that Nuclear Power today is not as cheap as it used to be. It may at best be competitive with more expensive sources such as oil and LNG.

Nuclear liability and insurance cover: Finally, Nuclear Power Economics may not be adequately covered without discussing the accident liability and insurance issues. Normal insurance policies do not provide coverage to third parties in case of nuclear accidents. Most nuclear nations have adopted national laws regarding protection to third party victims in case of nuclear accidents. This includes, the US, European Union countries, India and even China. The US has been pioneer in this respect with Price Anderson Act providing for a liability-insurance cover totaling 12.6 Billion USD covering all nuclear installations in the US. No government contribution is involved in it. Amended Vienna-IAEA convention and Joint Protocols provide the legal basis of similar arrangement in Europe. Nuclear liability has been limited to a total of USD 480 under Vienna protocols to be funded in some ratio with nuclear operators and respective governments. China is not a member of any international instrument in this respect, but has instituted national scheme covering nuclear liability and insurance. India has also introduced a similar arrangement under national legislation providing for an identical amount of USD 480 million USD. However, there is an innovative and yet controversial inclusion in the Indian Law, which passes on the liability to the vendor of the nuclear equipment, which in most cases are of foreign origin. All other national legislations hold nuclear operator to be solely responsible. Pakistan should also move in this direction and introduce national legislation along similar lines. Foreign Direct Investments would be discouraged if such legislation is not introduced. Also, the potential victims of nuclear accidents in the country should get some reassurance as, especially, when such controversial nuclear siting decisions such as of K-2 in Karachi are being taken. It may not affect nuclear power economics in a major way, as the actual premium payments per year may be around one million USD or so, if membership to an adequate international insurance pool is obtained.

Conclusion Pakistan has more than one reason to add nuclear power. There is a large nuclear establishment that has to be maintained and paid for. Lowest cost may not be the sole criterion to shape a country's energy options. Nuclear power would add diversity to its energy profile. It may be advisable to creatively explore ways and means of reducing the capital cost ala - Turkey-Russia deal ie permitting Chinese to own the nuclear plants and sell electricity at an affordable price. Easy financial terms like 4% interest and 20 years repayment period after commissioning may be explored, if there is no discount on EPC prices. It may not be a bad idea to flirt with Russians to get a competitive quote to be able to negotiate with Chinese. Also PAEC may be well advised to contain its program to 8000 MW only by 2030, keeping this rather discouraging price scenario in mind. They would be doing a job, if they achieve this target. In India and Pakistan, making large claims and unachievable targets for nuclear energy is part of patriotism. It may not be a bad idea to appear unpatriotic some time and be realistic.

42. Demystifying Nandipur Project Controversy

Nandipur power plant project, in 2015, attracted a lot of media attention. An analysis of the arguments that were put forth indicates that quite a few people were either confused or were ill informed. In this chapter, I would try to put together the relevant facts and figures and do some analyses so that a reasonably clear picture emerges and doubts and fears are removed from public specter. I have to clarify here that I have not been involved with Nandipur project and I can thus claim to present an unbiased and neutral account. Although some of the data may be drawn from government public documents, most of it is, however, from NEPRA, which can safely and hopefully be considered a source of unbiased and reliable information.

The impending questions that this chapter attempts to answer are: has CAPEX (capital expenditure) reached an impossibly high and unaffordable level? How does the CAPEX compare with other comparable plants? Was it more expedient to shun the project rather than revive it? Would there have been much greater loss in shunning the project as opposed to the additional cost that has been paid for in revival, as we would demonstrate later in these passages? What would be the cost of generation as compared to other comparable plants? And finally what are the prospects of running the plant without undue interruption?

Before delving into the answers of these questions, I'll first present a little perspective and project history. Nandipur Power Plant is situated in Gujranwala near an existing hydro power plant. It is a tri-fuel power plant able to run on Diesel, Furnace Oil and Natural gas. However, it works its best with natural gas generating 8.2% more power. Furnace Oil has to be pretreated before firing to meet the fuel specs requirement. A FO treatment plant has been provided along with FO storage facilities.

Table 53 Nandipur Turbine Specifications

Description	HSFO	Gas Fuel
Gross Capacity (Each Turbine)	95.4 MW	103.23 MW
Gross Capacity (Stream Turbine Unit)	138.9 MW	150.29 MW
Gross Capacity (Complex)	425 MW	460.00 MW
Auxiliary Consumption (Complex)	13.649 MW	12.328 MW
Net Capacity (Complex)	411.351 MW	447.672 MW

Source: NEPRA

According to original PC-I the original cost of the project was Rs 22.335 billion / USD 60 = $372.25 million (in 2008) which was revised upward to Rs 58.416 billion/ USD 105 = $556.34 million in 2014[9]. The EPC contract was awarded in 2008 to M/s Dongfang Electric of China by the PPP government. Machinery arrived in 2011 and kept lying at the Karachi seaport for about 2 years due to a legal issue In 2012, the present Minister of Water and Power Khawaja Asif Hussain filed a petition in the Supreme Court praying for a court order for saving the project. Supreme Court (tribunal) ordered implementation of the project.

Table 54 Nandipur Approved v Claimed Cost

Description	Approved $	Claimed $	Approved Rs.	Claimed Rs.
EPC	307	318	27,325	27,944
Escalation	N/A	65	N/A	6,726
Capex	363	535	32,946	49,955
IDC	82	146	7,373	14,324

Source: NEPRA

The present government with Khwaja Asif in power as the Minister of Water and Power revived the project after hectic negotiations with Dongfang involving diplomatic efforts. Dongfang was persuaded to resume the project at an additional cost/compensation of 64 million USD. First gas turbine was commissioned in May 2014 and was inaugurated by Prime Minister Nawaz Sharif. In July 2015, complete plant was commissioned with two gas turbines and one steam turbine. The plant only worked for a few days before it ceased operations.

---

[9] (All PC-1 costs are estimated only and it is quite common that PC-1 s are revised due to cost escalations due to late implementations, currency devaluation and shear under-estimates).

Nandipur Power Plant has a gross installed capacity of 425 MW on FO and 460 MW on gas. Its CAPEX now stands at 693 million USD, and not the 850 million USD as is being erroneously claimed. The project has been implemented over a period of seven years and the money has been spent at the prevailing exchange rates at different points in time during construction period, varying from Rs 67 to Rs 100 per USD. The project opponents are using one exchange rate in their calculations. One would note that this extra cost compares well with the compensation to contractor Dongfang that would have to be paid in case of terminating the contract and shunning the project. On a per MW basis, the revised current cost stands at 1,506 USD/KW, although high but not terribly high if compared with other plants; UCH-2 at 1472 USD/KW in 2008, Engro Qadirpur at 1010 USD/KW (2007) and Guddu at 1030 USD/kW (2011).

NEPRA issued a tariff for the project in 2015 at US cents 10.97 per unit on furnace oil out of which capacity charge is US cents 2.00 per unit and energy cost is US cents 8.97 per unit. On Natural Gas at prevailing gas prices, the Nandipur power tariff is expected to be Rs. 7.18 per kWh. Admittedly, these are slightly under-estimated figures as these are based on a reduced CAPEX allowed by NEPRA. If disallowed but actual costs are included, the capacity charge may go up by Rs. 1-1.5.

Is this too high? Not really. Today hydropower is costing in the range of Rs 8 to 10 per kWh as gone are the days of cheap hydroelectricity. Existing FO plants are generating power at a fuel charge alone of 9.8 cents (April 2015 around which Nandipur Tariff has been calculated by NEPRA).

Let us now discuss the extra cost that has been incurred. Nandipur tariff petition asked for a CAPEX of 535 Million USD and got only 363 Million USD from NEPRA in terms of allowed costs; evidently largest extra cost comes out to be the Interest during Construction (IDC) of 146 Million USD (21% of CAPEX). Normally in such projects, depending on interest rates, IDC should not exceed 10% of CAPEX. NEPRA did not allow 64 Million USD of extra costs paid to the contractor either, not because these were not genuine, but because NEPRA did not want the consumers to bear the consequence of the mistakes of governments. So put together, IDC plus extra cost paid to contractor comes out to be 210 Million USD. Add some escalation and currency devaluation as well. This is where the money appears to have gone.

Table 55 Comparison with other IPPs

O&M	Saif	Orient	Sapphire	Halmore	Average	Nandipur
HSD/RFO Foreign	0.50	0.33	0.49	0.50	0.46	0.53
HSD/RFO Local						0.18
Total Variable	0.50	0.33	0.49	0.50	0.46	0.71
Fixed O&M (Rs/kW/h)						
Local	0.10	0.17	0.10	0.10	0.12	0.09
Foreign	0.13	0.15	0.13	0.13	0.13	0.13
Total Fixed	0.23	0.32	0.23	0.23	0.25	0.22

Source: NEPRA

Table 56 Comparison with other Gas Plants

Power Plant	Guddu CCPP	Nandipur CCPP	UCH - II	Engro Qadirpur	RLNG (Upfront)
Capacity (MW)	747	425	386	226.2	400
Capacity Factor	60%	60%	60%	60%	92%
OPEX (Rs/KWh)	5.233	8.197	3.42	2.249	7.76
Capex (PKR Millions)	78,767	53,472	47,113	22,897	40,772
Capex (USD Million)	769.58	534.72	529.36	289.83	404.42
CAPEX USD/KW	1030.23	1515.86	1371.4	1281.32	1011.06
Levellised Tariff (Rs/kWh)	6.9	10.98	6.19	6.21	9.54
Gas Turbines (NOs)	2	3	3	1	-
GT (Capacity MW)	243	95.5		123	-
Steam Turbines	1	1	1	1	-
ST (Capacity MW)	261	138		110	-
Make of Turbine	GE	GE	GE	GE	-

Source: Compiled by author from various sources

Let us now come to the most important question. Will Nandipur plant work again? Is there something fundamentally wrong in it? There is no major problem with the project except minor issues like smaller capacity of FO treatment plant, which is being taken care of and should take at most six months. Ideally, the most befitting fuel for this plant is natural gas.

Admittedly, there has been some laxity and inefficiency in the appointment of O&M contractors. However, the contract was reportedly was awarded to the Chinese company, Hydro Electric Power System Engineering Company (HEPSEC) in February 2017. A total of four companies had participated in the bidding, but two were declared non-responsive. HEPSEC was confirmed the lowest bidder at a total cost of 185 million USD.

Another new phenomenon is that company boards are increasingly asserting their authority and independence and resisting instructions from the relevant ministries. A consensus seeking culture has yet to emerge, as board members have to be selected on merit rather than on influence and cronyism. CEOs are scared of NAB and accountability guard. Some of the finest and honest of CEOs and executives have suffered on travesties of justice in this regard. Thus decision-making has become slow, lame and even contentious. A new balance is being awaited out of the social turmoil that we are passing through.

Anyway, the sooner the issues are resolved, the better it would be. There is a monthly mark-up of about 3 million USD and in six months, it would add another 18 Million USD to the project cost. It is the IDC which is a bigger threat and not the perceptions or apprehensions of transparency issues. It should also be recognized that currency devaluation has also played a role in Rupee costs, otherwise, in dollar terms, the cost increase has not been the biggest issue, as is evident by a reasonable unit CAPEX rate as indicated earlier. The key lies in fast track running on RFO and eventually on natural gas or RLNG.

If at all, three governments are involved or culpable, if you will. Musharraf regime launched the project, PPP government partially implemented and damaged it and the present government retrieved and completed it. Sooner or not much later, it will generate electricity. Finally, if a lost project is retrieved, break-even should be taken as success.

About the Book

The author presents a thorough analysis of Pakistan's energy sector. Being an insider, he has had enough exposure to be able to point out issues and problems and has been able to develop a package of thorough recommendations. He has argued for better negotiations of CPEC energy sector's terms and has advised NEPRA to take steps to bring down tariff to a fair and affordable level but argues that Petroleum prices in Pakistan are fair and lowest in the region. The author is of the view that energy imports should be discouraged and argues in favour of local resource development and indigenization. In the back drop of remarkable recent reduction in Solar PV prices, he argues for a significant reappraisal of existing plans leading to a major initiative for inducting solar electricity in a distributed mode at about 50 locations. He has proposed alternatives and cheaper approaches with respect to Smart Meters, LPG distribution projects and RLNG terminals etc. Supporting coal, he has criticized avoidance of the required environmental controls and has demanded corrective steps in this respect. Overall, he presents an optimistic picture with critical evaluation of issues and problems. Laden with a lot of data and tables, the book should be a must reading for policy makers, stakeholders, academia and all those who have more than a passing interest in the subject.

About the Author

Syed Akhtar Ali is an eminent energy expert, an Engineer-Economist and consultant. His most recent assignment was as Member Energy, Planning Commission. He has been offering consulting services to public and private sector clients in the area of energy and environment, investments and tariff issues and has authored a number of books on these subjects. He was a visiting Professor of Energy at IoBM and has taught energy management to MBA students. He has held top management appointments in Pakistan's public and private sector. He was Research Fellow Energy at Harvard University's Kennedy School of Government. He is an author of eight books on various subjects such as energy, governance, political economy and resources. He has done consulting assignments in Energy and Environment Sector in the U.S., Europe, the Middle East, Africa and South Asia.

www.ingramcontent.com/pod-product-compliance
Lightning Source LLC
Chambersburg PA
CBHW082202220526
45470CB00010B/3016